COLLISION
EARTH!

COLLISION EARTH!

THE THREAT FROM OUTER SPACE

METEORITE AND COMET IMPACTS

Peter Grego

BLANDFORD

꣑DEC 6   1999

A Blandford Book

First published in the UK 1998 by Blandford

Cassell plc,
Wellington House
125 Strand
London WC2R 0BB

A Cassell Imprint

Distributed in the United States by Sterling Publishing Co., Inc.,
387 Park Avenue South, New York, NY 10016-8810

A Cataloguing-in-Publication Data entry for this title
is available from the British Library

ISBN 0-7137-2742-X

Designed by Linde Hardaker   3 9082 07614 5856

Printed in Great Britain by MPG Books Ltd, Bodmin, Cornwall

CONTENTS

ACKNOWLEDGEMENTS

I would like to thank Stuart Booth, Consultant Editor at Blandford, for having shown interest in this book and overseen its publication. Thanks also to Antonia Maxwell, Sarah Widdicombe and Jane Birch for their editorial work on the book.

My job as an author would have been impossible without the help, support and encouragement of Tina, my wife.

In researching illustrations, thanks are due to Janice Smith of the Geological Survey Canada, Maura O'Connor of the National Library of Australia, Ian Miller of AUSLIG, Mary Ann Hager of the Lunar and Planetary Institute (LPI), Irene Szewczuk and R. Lie of Sky Publishing, Robin Scagell of Galaxy Picture Library and Lesley Owen-Edwards of Unilever archives.

For the illustrations themselves, thanks to Virgil L. Sharpton of the LPI, David Roddy of the US Geological Survey, A. Scott Murrell of New Mexico State University, John Greco, the European Space Agency (ESA) and the US National Aeronautics and Space Administration (NASA).

I would like to thank librarians Bernard Ruffell and John Skidmore of the Birmingham Astronomical Society for all their help, and thanks to Major Jonathan Tate of Spaceguard UK, Rob Childs, Paul Stephens and Phil Lewis for their expert assistance.

INTRODUCTION

The city of Birmingham, England: 17 April 1984, 01:18 UT. A cool, crisp and clear night. A reasonably dark night, too, considering the orange sky-glow around the horizon created by countless city lights and the presence of a Moon just past full, hanging low above the southern cityscape in the constellation of Libra.

The amateur astronomer was pleased with the results of his session so far. That evening he had already observed the planet Saturn, to the naked eye a bright yellow star lying just 1 degree above the Moon. Through his telescope, Saturn's mighty system of rings appeared wide open and confidently advertised itself as one of the solar system's most fabulous sights. He imagined he was looking through the porthole of his very own spacecraft, going up close to Saturn and examining the rings, just as the spaceprobe Voyager 2 had done less than three years previously.

Voyager proved the rings to be vast fields of debris – countless chunks of ice, rocks and dust grains, each particle in its own individual orbit around the planet within the narrow ring plane. The Voyager probes had been fortunate not to suffer damage from larger ring particles as the spacecraft sped past Saturn on their paths outward from the Sun. At a relative closing velocity of a staggering 13km per second, the impact of a single sand-grain-sized dust particle could have blasted quite a hole in the probe, with the potential to disable or even destroy it outright. As it happened, the Voyager did hit some outer ring particles, but these were microscopic and only detected by the signal interference caused by the impacting particles vaporizing into a temporary cloud of plasma, which was picked up by the probes' sensitive radio antennae.

The young astronomer pondered: if, somehow, Saturn suddenly vanished, then these countless particles might begin to spiral towards the Sun and become the solar system's newest shoal of meteoroids. For all Saturn's enthralling beauty, the astronomer had managed to turn his telescope away from the ringed planet – tempted though he had been to gaze at the object for a while longer – and had assigned his attention to the Moon. For the past hour or so he had been engaged in a telescopic study of the lunar crater Condorcet, some 74km wide, which may have been blasted out of the Moon's rocky crust thousands of millions of years ago, when an asteroid several kilometres in diameter impacted on to the lunar surface at high velocity. He had made a careful drawing of the crater, showing the intricate contortions of its walls and the unusual structure on the floor. While sketching Condorcet, the astronomer mused about asteroids and the effects of cosmic impacts upon the Moon and other planetary bodies. What a sight those innumerable lunar impacts must have been – rather a pity that only a handful of small lunar impact craters have been formed in the couple of million years since mankind appeared on this planet, and of these maybe just one has occurred in all of recorded human history.

Now happy with his sketch, the astronomer prepared to make an observation of Mars, which to the naked eye appeared as a bright orange star just a few degrees east of the Moon. Swinging his telescope around, he suddenly became aware that the scene around him had brightened considerably. Almost directly overhead, in the tail of the Great Bear, a brilliant light had materialized in the sky. The hairs on the nape of his neck stood up as, in utter silence, the light flared brighter and shot towards the northeast, suddenly bursting and breaking up into a dozen flaming yellow pieces as it went. A couple of seconds later, as the fragments sped like a formation of burning arrows through the constellation of Draco and towards Cygnus, the objects faded out, leaving a distinctly bluish trail in their wake which lasted for a few more seconds in the still night air. At first, the astronomer imagined that he had witnessed the mid-air explosion of an aircraft – but there had been no noise, and a moment later he concluded that the magnificent object had been a brilliant fireball, the visitation of a small rock from the depths of space which had exploded during its superheated descent through the Earth's atmosphere.

Only three weeks earlier, the amateur astronomer had spent two consecutive nights keeping a vigil under the cold night sky in the hope of seeing a few members of an annual shower of shooting stars known as the Virginid meteors, so-called because they appear to radiate from a point in the constellation of Virgo. The Virginids are in fact one of the least impressive of the year's many meteor displays, with low rates of activity occurring between February and mid-May. He had managed to record only four Virginid meteors in a total of several hours, each of them having been rather faint and unspectacular. Still, he had considered the sight well worth staying up for, even though his school work had suffered slightly on the mornings after. The fireball he had just witnessed seemed to be some sort of reward for his Virginid efforts – although who or what considered the astronomer a worthy recipient of such a cosmic gratuity remains unknown.

Invigorated by the sight, the astronomer remained outside for another hour, and, while making his planned observation of Mars and another lunar drawing, he paused from time to time to survey the heavens briefly in the hope that another fireball might appear. The second lunar feature he chose to draw – a chain of small craters to the east of the large crater Langrenus – was very possibly excavated (long before dinosaurs ruled the Earth) from the Moon's surface by the simultaneous impacts of a dozen or so meteoroids, much like the fragmenting fireball he had just seen, only billions of times more massive.

Such was the night on which I had my first encounter with an extremely bright meteor – an impressive fireball which fell in the morning skies over the English Midlands. I like to think that some of the original object, maybe a few pieces the size of a peanut, managed to survive their fiery descent through the atmosphere to land in some country field in the northeast of the country.

Yet this event, so spectacular as to be permanently imprinted in my memory, could pale into insignificance when compared with a meteor storm – a fall of shooting stars which takes place in such rapid succession that even experienced astronomers admit to being rendered completely in awe of the spectacle, if not a little fearful of it. The last true meteor storm took place a generation ago, in November 1966, a month before my first birthday.

As I write, a meteor storm of incredible proportions might be brewing on the interplanetary horizon. There is a distinct possibility that a new generation of sky-watchers will be treated to one of astronomy's most glorious sights – a frenetic display of cosmic fireworks predicted to occur on the morning of 18 November 1999, heralding the eve of the new millennium. Far into the twenty-first century, the year 1999 may come to be remembered as the year of the great twentieth-century meteor storm.

This spectacular meteor activity will appear to radiate from an area in the constellation of Leo over a period of a few glorious hours. Astronomers predict that the chances of a significant meteor storm (perhaps even more than one storm) will actually be greatest between the years 1998 and 2000, but the 1999 event is likely to be the best of the bunch and it will be visible in British skies. By studying historical meteor accounts carefully and combining these with up-to-date scientific data, many astronomers deduce that 1999 looks to be the year in which a Leonid storm is most probable.

The Leonid storm has been on the cards for more than a century, for astronomers have long known about the special conditions required for meteor showers and associated meteor storms to be visible in the Earth's atmosphere. Every year, the Earth encounters more than a dozen rivers of meteoroidal debris which have been deposited in space over thousands of years by comets, whose nuclei are masses of compacted ices, rock and dust a few kilometres across. In effect, comets are massive dirty snowballs, whose ices sublimate (change from solid ice to gas) when they approach the Sun and gradually warm up.

Our planet ploughs into these meteoroid streams at certain well established dates every year. The meteoroidal dust particles quickly burn up as friction with the Earth's atmosphere heats them, leaving a glowing trail in the sky called a meteor. Occasionally the Earth moves through a particularly rich part of a meteor stream, and at these times the number of meteors seen will be greater than usual.

Every November, without fail, we see meteors apparently emanating from the constellation of Leo. Each year, the Leonid meteor shower takes place between 15 and 19 November, with a peak of activity on the morning of 17 November, when the number of meteors you can normally expect to see is about ten every hour (this figure varies a little from year to year). The Leonid stream was deposited in space by a comet called Tempel-Tuttle (named after the pair of astronomers who independently discovered the comet in 1865–6), and this comet orbits the Sun once every 33 years. In the immediate wake of comet Tempel-Tuttle there is a particularly condensed swarm of meteoroids which follows the comet in its 33 year orbit. Every 33 years the Earth passes very close to, or even through, this Leonid meteoroid swarm when it is in our vicinity. These are the times when great Leonid meteor storms take place, and they have been observed with wonder by many generations of sky-watchers.

Accounts of Leonid storms go back as far as the year AD 902, to the days when superstition governed many people's perception of events in the night sky. To an uneducated eye and a superstitious mind, such a meteoric phenomenon would be terrifying to behold – a signal, perhaps, of the wrath of the gods and a portent of some terrestrial disaster. It is probably no exaggeration to claim that most people who observed these incredible, unexpected heavenly eruptions actually feared for their lives and imagined that the end of the world was nigh. The first modern account of the Leonid storm appears in 1799, and the years 1833, 1866 and 1966 were each blessed with fantastic Leonid displays.

Just a thin layer of atmosphere separates us on the ground from the harshness of the interplanetary vacuum. Just as the ozone layer protects our skin and eyes from the Sun's harmful ultraviolet radiation, the gaseous envelope of the Earth's atmosphere is currently our only defence against incoming objects. If we weren't shielded in this manner then meteoroids would hit the ground faster than bullets, peppering the landscape (and any unfortunate people) with a multitude of impact pits.

The Moon has no atmosphere at all, and a look at its surface through a telescope or a glance at a lunar photograph will show you exactly what our own planet would look like if it had no atmosphere – a world packed with craters. Meteoroids are the Moon's main cause of erosion, as the lunar surface is continually sandblasted by high-speed interplanetary debris. The bright flash of a meteor in the Earth's skies should reassure the observer rather than inspire fear, because as long as we have an atmosphere to protect us from the majority of cosmic impacts, we can all gaze into the heavens and marvel at meteor showers and meteor storms without too much cause for concern.

However, by delving further afield to examine interplanetary debris and cosmic impact – from the demise of the dinosaurs to the dangers faced by spacecraft in Earth orbit – I hope in this book to remind the reader that those innocent-looking meteors that flash across the sky lie at one end of a spectrum of celestial phenomena which can shape entire worlds and influence the minds of their inhabitants. Perhaps it will take a major meteor storm at the end of the twentieth century finally to convince governments to consider the threat of cosmic impact seriously, and to put some vital funding into the area. A word of warning to keraunothnetophobes (those with an irrational fear of being struck from above) – this book contains the stuff of your worst nightmares!

Whatever happens in November 1999 and the years clustered around that date, I hope this book will encourage the reader to enjoy a facet of astronomy which is entirely free of charge and requires absolutely no optical aid. The patient sky-watcher has a cast-iron guarantee of success, because meteors can be seen with little effort during each of the dozen or more annual shower periods outlined in these pages.

With a little luck, something truly wonderful may be seen in the Earth's skies at the end of the twentieth century – a majestic star-storm the like of which may not be repeated in our lifetimes.

Peter Grego
West Bromwich, September 1997

1

COMET
TALES

THE HIDDEN REALM OF COMETS

When the German astronomer Johannes Kepler (1571–1630) claimed that there are more comets in the heavens than there are fish in the sea, he was merely making a wild guess to impress his audience with an idea of the vastness of the solar system. Kepler may have thought he was exaggerating, but in fact his estimation was right. Far beyond the planets – so far away that the Sun is reduced to a bright, star-like point – a wide zone of space, populated by countless icy cometary nuclei, marks the very outer fringes of the solar system. The size and shape of this comet zone, members of which have never been observed, remains speculative.

According to the comet expert Donald Yeomans (see Bibliography), the zone is a flattened disc just beyond the planetary region, extending to 10,000 Astronomical Units from the Sun (1AU is the average distance between Earth and Sun, some 158 million km). The region then broadens into a spherical outer zone, with a rugby-ball-shaped outer shell pointing towards the galactic centre, measuring 200,000 x 160,000AU. Containing an estimated 100 billion comets, this zone has never been observed directly, nor do we currently have much prospect of detecting individual comets at that enormous distance, although advances in imaging technology may possibly change this in the years to come.

In 1950, the Dutch astronomer Jan Oort arrived at the first feasible theory regarding the origins of the comet zone, and in his honour the zone has been named the Oort Cloud. Oort once speculated that the cloud may be the debris of a Mars-sized planet which was ripped apart by gravitational forces early in the history of the solar system. Most of the fragments were lost to the solar system forever, but about 3 per cent of them were flung into orbits taking them as far as 25,000 to 200,000AU from the Sun, according to Oort's original hypothesis. Oort postulated that the solar system could, theoretically, have held on to 100 billion fragments of the original exploding planet, many of which became cometary nuclei, each with a mass of 10 billion tonnes.

Anatomy of a Comet

It is surprising what a lump of dirty ice just a few kilometres across can do. Much further away than the orbit of Neptune, cometary nuclei are too small and faint to be visible through optical telescopes. At present we can only dream of seeing the individual members of the Oort Cloud, which possibly extends nearly halfway to the nearest star. But when a comet's orbit around the Sun takes it into the inner reaches of the solar system, it begins to respond to the increased heat and energy received from the Sun.

The various ices on the surface of the comet's nucleus – substances like water, ammonia, methane, nitrogen, carbon monoxide and carbon dioxide ice – start to change from their solid icy state to vapour. This process, known as sublimation, also allows tiny dust grains to escape from the surface.

Gases released from the surface of the comet's nucleus are bombarded by the solar wind, a 500km per second stream of electrons and protons that continually expands away from the Sun into the depths of inter-planetary space. The comet's gases are said to ionize as they pick up some of this electric charge from the solar wind. These ionized gases form a cloud around the comet's nucleus which is brushed away by the solar wind, and the nucleus begins to grow a long ion tail flowing away from the direction of the Sun.

Small dust grains are released into space as the comet heats, and the pressure of the solar radiation causes a separate dust tail to develop alongside the gaseous ion tail. It is the release of these dust particles that provides the stuff of meteor streams. In 1983, the Infrared Astronomical Satellite (IRAS) actually detected more than 100 faint trails of meteor-oidal dust within the solar system, and identified distinct trails associated with the comets Encke, Tempel 2 and Gunn. Comet Encke's dust trail (from which is derived November's Taurid meteor shower) was measured on IRAS images to be 300 million km long – equal to the diameter of the Earth's orbit around the Sun.

Some of the more active comets become really bright objects, and if conditions are right they can be impressive celestial spectacles for a period of time. Nobody in the northern hemisphere will ever forget the glorious Comet Hale-Bopp which appeared in the northwestern evening skies during the first half of 1997, with its bright, star-like nucleus, curved yellow dust tail and straight bluish ion tail. Other comets – indeed, the majority of comets observed – never reach naked-eye visibility and remain dim objects visible only through binoculars and telescopes.

Martian Connections

Streams of meteoroids are produced by the dust released by cometary nuclei as their ices sublimate. The first scientist to prove that comets produce the meteoroid streams was the Italian astronomer Giovanni Schiaparelli (1835–1910) in 1862.

To digress briefly, Schiaparelli's name will forever be linked with his more famous discovery, in 1877, of linear markings stretching for tens of thousands of kilometres across the red disc of Mars. He named these faint features *canali* ('channels'), but was careful not to link their origin with speculation about possible life on the red planet, as such theorizing was rife in the late nineteenth century. Schiaparelli thought the markings he saw through his 220mm refractor might represent large fissures or some other natural large-scale geological process of the planet's crust.

But the Martian discovery was announced to a world hungry for sensation, especially in the exciting field of astronomy. The real meaning of the word *canali* was ignored by the majority, and associated instead with canals. This was an era that was witnessing ever more impressive feats of engineering. From April 1859 to November 1869, thousands had laboured in carving out the 163km of the Suez Canal, providing a vital navigable link between the Mediterranean Sea and the Gulf of Suez at a cost of $100 million.

The Martian 'canals' were proclaimed to be sensational observational evidence of vast waterway systems hewn out of Mars' red deserts by an advanced race of Martian beings. Estimated total cost to the Martians in 1877 US dollars (given a total length of a 250,000km, although even this is an underestimate): $150 billion. This amounts to just six times the cost of the Earthlings' own Apollo Moon-landing programme a century later. The dusky lines observed through the telescope by dozens of subsequent eagle-eyed astronomers (both amateur and professional) in the ensuing decades of the late nineteenth and early twentieth century were envisioned to be fertile areas bordering the canals, tracts of vegetation flourishing out of the freshly irrigated Martian sands.

Intelligent Martians have never existed, nor have the canals of Mars. Many of the features seen were misinterpretations of unresolved dusky alignments and boundaries defining areas of contrast – pure illusions. Spaceprobes have shown the surface of Mars in all its glory. There is one enormous fault valley, the Mariner Valley (Vallis Marineris) which dwarfs the Earth's Grand Canyon, and this feature may actually have been glimpsed by nineteenth-century telescopic observers. There are also many smaller channels on the Martian surface which appear to have been cut by running water. Mars' running water has long since vanished; much of it is tied up in layers of permafrost below the surface and in the polar ice caps.

One phenomenon certainly shared by the Earth and its neighbour Mars is the appearance of meteors in their skies, only above Mars they will burn up much closer to the planet's surface, at a height of around 20km. The meteors which will be seen in Martian skies by the first human settlers in the middle of the twenty-first century will not come from the same streams as our own, for Mars orbits in a markedly eccentric path, ranging from 263 million km at aphelion (furthest from the Sun) to 218 million km at perihelion (nearest to the Sun). The Earth has a near-circular orbit around the Sun, averaging 158 million km. Mars sometimes approaches Earth to 60 million km at the so-called perihelic oppositions, which occur every 17 years or so. Mars has its own unique meteor showers, which may or may not be as active as those we see from the Earth.

COMETS: THE ORIGIN OF METEOR SHOWERS

It is a pity that Schiaparelli will be remembered mainly for his non-existent Martian canals in the minds of generations of lay-people and astronomers – one of those cruel twists of fate which tend to distort the picture of the lives and works of scientists. But Schiaparelli made a much more important discovery in an altogether different astronomical speciality – comets and meteors.

Schiaparelli's most important work, much of it conducted while he was settling into his new job as Director of the Brera Observatory in Milan, Italy, involved the relationship between meteor showers and comets. The great astronomer calculated that orbital elements of the annual Perseid shower in August (based on a parabolic 'open-ended' orbit) matched the elements of a comet discovered in July 1862 (the third comet to have been discovered that year, then referred to as Comet 1862 III). Schiaparelli demonstrated mathematically that the comet and the members of the Perseid meteor shower, crossing the Earth's orbit in the same place, have directions of travel very similar to one another and follow the same course through space.

Comet 1862 III is now known as Comet Swift-Tuttle after its discoverers, the American astronomers Lewis Swift and Horace Tuttle. In August and September 1862, Swift-Tuttle brightened and became a prominent second magnitude object in the evening sky, with a tail 10 degrees long. Schiaparelli pointed out that the Perseid stream is likely to be composed of trains of dusty debris flowing along the very elliptical orbit of Swift-Tuttle.

Most astronomers expected Swift-Tuttle to make its return appearance in 1982, in accordance with the predicted 120 year orbital period. However, the comet did not reveal itself until September 1992, when it was picked up through giant 25 x 150 binoculars as a faint 11th magnitude patch by the Japanese amateur astronomer Tsuruhiko Kiuchi. Swift-Tuttle was never a very large comet, but it was easily visible in binoculars and went on to develop a short tail (around 1 degree in length if seen from light-polluted urban areas) as it coasted across the evening skies towards its closest encounter with the Sun at perihelion.

Good, more or less reliable displays of at least 50 meteors per hour may be observed at the shower's maximum on most years, meaning that the material of the Perseid stream deposited by Comet Swift-Tuttle is very evenly spread through space. Because of this fact, it has been asserted that the shower is very ancient. Over a long period of time, the billions of particles with their slightly different speeds have managed to disperse themselves around the entire elliptical orbit (which goes well beyond Pluto's orbit) of the cometary progenitor. Simon Newcomb (1835–1909) stated that the evenly spread Perseids may be 20 times older than the Leonids – making the Perseid shower over 60,000 years old. The first European records of the Perseids date back to AD 830.

Biela's Comet and the Andromedid Meteor Shower

On 28 February 1826, at the garrisoned hamlet of Josefov in Bohemia, a Czech-born Austrian soldier and part-time amateur astronomer, Captain Wilhelm von Biela (1782–1856), discovered a comet which was to astonish the scientific world. As the weeks went by, and news of the discovery spread, Biela's Comet became as bright as the third magnitude and was easily visible to the naked eye. The new comet was found to match a comet discovered by Charles Messier amid the stars of Eridanus in 1772, and another seen in 1805; it soon dawned upon astronomers that all three comets, each with a period of 6 years 8 months, were undoubtedly the same object.

Biela's Comet came and went through our planetary neighbourhood again in 1832–3, actually having been recovered in September 1832 by the famous English astronomer John Herschel. The comet's apparition in 1839 went unobserved because of the poor position of the Earth in relation to it, but it made a good showing at its next apparition in 1845. Observers noted a peculiar tail streaming from the comet's head in the general direction of the Sun – the first to notice this was the American astronomer Edward Herrick, who also detected a knot-like structure in this so-called 'anti-tail' which was more brilliant and more condensed than any other part of the coma.

In early January 1846, Lieutenant Matthew Maury at the US Naval Observatory was astonished to discover that the comet had split into two distinct pieces connected by a faint hazy trail of dust and gas. The two components – each with their own fuzzy coma and small tail – were at first unequal in brilliance, but over the next month the smaller companion brightened and became the other's equal. It then faded, and in March it disappeared completely, while its companion continued to be visible for another month or so. Professor Hubbard calculated the two objects to be around 340,000km apart.

On Comet Biela's return in 1852, the pair had moved apart from each other to a distance of 2.6 million km. The two companions took turns at being the brighter, and it was not possible to determine which of them was the main body. The pair plunged into the depths of interplanetary space in September 1852 and have never been seen since. As the positions of the Earth and the comet were unfavourable in 1859, nobody was surprised that the object was not recovered, despite an intensive search. However, in the 1865–6 season Biela's Comet should have been easy to observe, but no trace of it was glimpsed.

Through the pioneering work of people like Schiaparelli, astronomers were by now aware of the comet-meteor connection, and many predicted that in November of 1872 a shower would occur when the Earth was in the vicinity of Biela's orbital path. In 1867, Edmund Weiss of Vienna stated that, according to his calculations, a grand Andromedid meteor shower would take place on 28 November 1872. Weiss' prediction was a day out. On 27 November a magnificent meteoric display took place, with its radiant in the constellation of Andromeda, on the very same day that the Earth intersected the path of Biela's Comet. If Biela was still out there, it would have been at that point in its orbit two months previously. Although the shower was by no means as spectacular as the Leonid storm of 1866, many who saw it were of the opinion that it was better than any other they had ever seen, Leonids excepted. E. J. Lowe at Highfield Observatory near Nottingham, England, estimated that he observed no less than 58,660 meteors between 17:50 and 22:50 UT. He remarked:

> The striking feature (more especially in the earlier portion of the display) was the extreme smallness of the great portion of the meteors, not one in ten being equal to a star of the third magnitude, and many were as minute as the smallest visible stars and might aptly be called meteor dust.

The next Andromedid meteor spectacular was seen 13 years later, on 27 November 1885, a display which ranked as being equal to, if not better than, the shower of 1872. The whole event was observed in clear skies by C. M. Vaison in France, who, writing in the 4 December issue of *The English Mechanic and World of Science*, said:

> Last night, 27th inst, we had here a most beautiful display of meteors, radiating from the neighbourhood of Gamma Andromedae. Hundreds of people went out of their houses to contemplate with very diverse feelings this beautiful phenomenon. Fear however was uppermost as some people, men and women, came to me to ask in voices altered by emotion, what all this meant; whether it was a sign of inundation, of war or pestilence, or whether we were all going to be squashed or burnt

by the stars falling upon us. They were soon reassured, I am glad to say, by a few words of mine and many were their expressions of gratitude as they went away.

Not having seen at the time the circular of Lord Crawford I did not expect this display myself. Soon after sunset I was observing Beta Delphini when I was surprised to see shooting stars crossing the field of my telescope. Other shooting stars, seen sideways, made me look up, and there I saw many large ones crossing various parts of the sky, their direction being at that time from east to west. No star was yet visible to the naked eye. As darkness came on, the whole expanse was crossed by these flying brands. Some were just blazing away as if gunpowder had exploded at that particular spot of the sky; others crossed the zenith for a short distance and many others bright and faint were falling from various heights in the heavens all around down to the horizon, but all seemed to have the same focus. Many of the larger meteors left long trails after them, lasting ten to twelve seconds, and then melting slowly away.... At times the meteors were so numerous that it seemed as if all the stars were moving.

The sky was perfectly clear from zenith to horizon all around; a strong dew was forming and we had a splendid glow at sunset. There is no doubt that this display began before sunset but by ten o'clock it was evident that we had nearly crossed the path of these meteors.

On the whole, British skies were more cloudy. Professor Robert Grant at the Glasgow Observatory, Scotland, was lucky – he had clear skies that night. Grant also had the good fortune to have seen the previous Andromedid shower of 1872 and the Leonid star-storm of 1866. Being in a unique position to compare all three events, he wrote:

It has been my good fortune to have seen from this observatory the great meteor shower of 13 November 1866, and also that of 27 November 1872, in both instances under exceptionally favourable circumstances, and at the time of the recent shower I was naturally led to institute a comparison between the brilliant apparition then visible and the two preceding displays. The shower of the Leonids appeared to me to be beyond comparison the grandest of the three apparitions, both in respect to the abundance and the magnitude of the meteors. On the other hand, the recent apparition of the Andromedids offered a most striking resemblance in every respect to its brilliant predecessor of 1872.

It was during the great Andromedid shower of 1885 that the first photograph of a meteor was secured on a plate taken by Ladislaus Weinek, Director of Praha Observatory at Klementinium in the Czech Republic. 1885 was notable for the discovery of the first extragalactic supernova: the light of an exploding star in the Andromeda Galaxy took 2.5 million years to reach the eyes of Victorian astronomers and was almost bright enough to be visible with the naked eye. At the time of the Andromedid shower the supernova had faded to magnitude 13 and was visible only in moderate-sized telescopes. It was an amazing coincidence that the two events were centred on the one constellation.

Another coincidence was the fall of a meteorite, on the night of the great Andromedid shower, at Conception Ranch, 13km east of Mazapil in Mexico. At the time, the meteorite, a lump of iron weighing 4kg, was just the ninth iron meteorite seen to fall, and the only one known to have fallen during the peak of a meteor shower. However, the object's initial path through the sky was not observed, so its radiant point is unknown. It is unlikely (although by no means impossible) that the Mazapil meteorite originated in the Andromedid stream. Eulogio Mijares, the rancher who recovered the object from a crater 30cm deep, described the fall (quoted by John G. Burke – see Bibliography):

> It was about nine o'clock in the evening when I went to the corral to feed certain horses. Suddenly I heard a loud sizzling noise exactly as though something red-hot was being plunged into water, and almost instantly there followed a somewhat loud thud. At once the corral was covered with a phosphorescent light and suspended in the air were small luminous sparks as though from a rocket. I had not recovered from my surprise when I saw this luminous air disappear, and there remained on the ground only such a light as made when a match is rubbed.
>
> A number of people from the neighbouring houses came running towards me, and they assisted me to quieten the horses which had become very much excited. We all asked each other what could be the matter, and we were afraid to walk in the corral for fear of getting burned. When in a few moments we had recovered from our surprise we saw the phosphorescent light disappear, little by little, and when we had brought lights to look for the cause we found a hole in the ground and in it a ball of light. We retired to a distance fearing it would explode and harm us. Looking up to the sky we saw from time to time exhalations of stars which soon went out but without noise.
>
> We returned after a little while and found in the hole a hot stone which we could barely handle, which on the next day looked like a piece of iron. All night it rained stars but we saw none fall to the ground as they seemed to be extinguished while still very high up.

Immediately following the 1885 Andromedid shower, one Professor Wilhelm Klinkerfues of Göttingen in Germany, realizing that the comet responsible for the dramatic meteoric outburst might be nearby, excitedly sent a telegram to his acquaintance Norman Pogson in Madras, India. The telegram (transmitted via numerous relay stations through

Russia and taking more than an hour and a half to reach its destination) stated: 'Biela touched Earth last night. Search near Theta Centauri.'

Pogson was able to make a search only two days later after the weather had improved. On 2 December he discovered an apparent cometary object which lagged behind the calculated position of Biela's Comet by two months. Pogson reported seeing the comet again the following evening, but as his observations were not corroborated by reliable sources, and the positions he gave were wide of the mark, the new comet could not be reconciled with either component of Biela's Comet or in any way implied to be a portion of Biela's nucleus which had mysteriously been delayed in its orbit.

Celestial mechanics often confound common sense. If an object in orbit around the sun is slowed down – say it encounters a barrage of debris in its path – then it will move towards the Sun, spiralling closer to it and actually speeding up its orbital period. Another method of applying orbital brakes is to fire retro-rockets, which provide a burst of energy directed ahead of the spacecraft's motion. All manned spacecraft do this to return to the Earth from their various orbits. A comet could apply its own unique retro-rockets by erupting violently from its volatile icy nucleus on its preceding hemisphere. But any eruption of Biela's Comet would have had to have been pretty energetic to have given rise to such a substantial delay in arrival – besides, there is the added problem of a consequent dramatic change in its orbit, which would have altered its observed position considerably.

Records of the Andromedid meteor shower long precede the splitting-up of Biela's Comet, having been traced back to AD 541. There was a very good Andromedid shower seen from Russia in 1741, and careful observations of the 1798 shower enabled two German students, Heinrich Brandes and Johann Benzenberg, to make the first calculation of a meteor's height above the Earth. To do this, Brandes and Benzenberg set up observing stations several kilometres from each other, and simultaneously noted the positions of the tracks left by meteors observed between September and November. Because the distance between the observers was known accurately, and the observed locations of the meteors could be measured against the starry background, all that was needed was simple mathematics (a technique called parallax) to determine the real height of the meteors above the ground. Noting where the meteors appeared to become extinguished, most meteors were shown to have faded out at heights between 10km and 214km, giving a mean disappearance height of 89km. Meteors were proven to be an atmospheric phenomenon likely to result from the infall of small extraterrestrial particles.

Perturbations by Jupiter have diverted the Andromedid stream away from its intersection with the Earth's orbit. The once busy Andromedid shower diminished in intensity and had more or less faded into obscurity by the middle of the twentieth century, although strangely some textbooks still continued to advertise the meteoric offspring of Biela's Comet as an annual shower of some relevance.

LATE TWENTIETH-CENTURY COMETS

The 1980s and 1990s saw many comets make their way across the skies, but only a few of them were at all spectacular in terms of their apparent brightness. The news media were thoroughly guilty of hyping most cometary apparitions – notably Halley in 1985-6, a fact which led to a great deal of disappointment among members of the general public, who were expecting much more of a visual spectacle.

Iras-Araki-Alcock (1983 D)

The comet was first reported from Peterborough in the UK by veteran amateur comet hunter George Alcock using his 15 x 80 binoculars. Its discovery is jointly shared with Japanese Genichi Araki and the orbiting IRAS. Iras-Araki-Alcock was well publicized in both the general and astronomical media as it made a nice showing in the first half of May 1983. Although it wasn't a truly spectacular sight, the comet was visible to the naked eye with little difficulty, even from light-polluted urban areas. Through binoculars and telescopes it appeared as a diffuse, elongated patch of light some 2 degrees across with a tiny, bright nuclear condensation.

Iras-Araki-Alcock was well placed for observation in the northern sky, and a spell of good weather conditions during the apparition made it a very well-observed comet. It was a swift-moving object, passing through the constellations of Draco, Ursa Minor, Ursa Major, Lynx, Cancer and Hydra in the space of just five days. At high magnifications, the motion of the comet against the starry background could actually be discerned in real-time, a fact that astonished most astronomers previously used to heavenly sights which usually appear static.

Halley's Comet

This most famous cometary visitor from the depths of interplanetary space was recovered in 1982 on CCD (a light-sensitive electronic device) images made with the almost equally famous 5m reflector on Mount Palomar in California. When recovered, Halley's Comet was exceptionally faint and nearly five years away from becoming bright enough to be visible in average amateur telescopes. Like many other amateur astronomers in the northern hemisphere, I followed Halley from early November 1985 to April 1986, with a break in early 1986 as the comet rounded the Sun at perihelion and was temporarily lost to view.

There was a tremendous amount of media hype surrounding Halley, even though astronomers had long predicted that the comet would not be a brilliant sight from the northern hemisphere. In the event, the press turned on Halley when it became clear that the cometary visitor hadn't been reading its own publicity material! Typical newspaper reports accused Halley of being a 'crummy fuzzball' or a 'soggy squib'.

An International Halley Watch was organized as a collection point for observations and to disseminate news about the comet. Professional astronomers lavished their full attention on it and employed the latest imaging techniques to squeeze as much information as possible out of Halley's precious light.

A small armada of unmanned spacecraft constructed by Russia, Europe and Japan was sent to intercept Halley's Comet. Hopes that America would join in the deep space investigations were not realized, as budgetary cutbacks meant that NASA (the National Aeronautics and Space Administration) hadn't enough money to fund its ambitious dedicated Halley mission. In March 1986, the two Soviet VeGa probes and the ESA (European Space Agency) Giotto craft passed through Halley's coma. VeGas 1 and 2 passed within 8,890km and 8,030km and of Halley's nucleus respectively, taking scores of low-resolution images as they sped by. Giotto approached the nucleus to within just 605km and took many stunning high-resolution images.

One problem the Giotto engineers had to attempt to minimize was the damage caused by impacting dust grains as the probe approached the comet at an incredible relative speed of 68km per second – 50 times the speed of a rifle bullet. It was decided to give the probe a double dust bumper on its leading end and spin it around its long axis once every four seconds in order to stabilize it in flight. Special 'nutation dampers' could be activated to cancel out any undue wobble in the craft's motion which might be caused by a small impact.

Giotto's multicolour camera was mounted at its forward end, peeking at a shatterproof metal mirror through a cutaway in the dust bumpers. The spin of the craft meant that the images were made up of circular sweeps which took in a section of the coma and were aligned by onboard computer to include the brightest part of the comet's nucleus on each sweep. An ingenious experiment was devised to detect the size and locations of impacts of small dust grains on the dust bumper. Some scientists gave the probe a one-in-ten chance of being destroyed completely as it headed towards Halley's nucleus – not exactly a kamikaze venture, more a fighting chance of survival in a single-shot game of Russian Roulette.

Because Giotto's camera kept the brightest part of the scene in its field of view, images of the dark nucleus which gave rise to these bright jets of dust and gas were visible only when the craft was some distance from the nucleus. The last images containing views of the dark nucleus were taken about 50 seconds prior to closest approach, a distance of some 3,500km. Further in, the images show just the bright nuclear jets and a certain amount of fine structure within them.

Millions watched the final phase of the Giotto approach near Halley's heart as the event was covered live on television. Viewers watched excitedly as monitors flashed a rapid succession of false-colour images transmitted by the large Earth-pointing parabolic dish attached to the rear of the distant spacecraft. In all, the probe returned more than 2,000 pictures of Halley.

The first dust particle to impact on Giotto's bumper was felt when the probe was some 287,000km from the nucleus. Between a distance of 150,000km and 90,000km there was a steady pitter-patter of small impacts, after which the rates fell until the probe was 70,000km from the nucleus, when activity began to increase steadily. The amount of dust detected in these outlying regions of the coma had been predicted to be far higher.

But just as astronomers began to think that Halley's coma was a relatively dust-scarce zone, 20 seconds prior to the probe's closest approach the dust-impact rate went off the measurable scale, with the thickest zone of dust encountered in the last minute before closest approach. More than 100 of the larger-sized dust particles in the last stages of approach managed to bump Giotto into a temporarily unrecoverable attitude. The last images secured by Giotto were taken at a distance of 1,350km from Halley's nucleus and showed a fan-shaped jet region about 15km across.

Halley's nucleus was found to be a dark, elongated, potato-shaped body around 15km long. Later image enhancements have revealed a degree of topographic form, including scalloped ridges, hills and several craters. Some of the craters may be impact scars, while others are undoubtedly areas excavated when pockets of volatiles under Halley's crust have exploded.

In February 1991, when the comet was outside Saturn's orbit, more than 2 billion km from the Sun and heading for a few decades of peaceful frozen semi-retirement in the outer solar system, astronomers were astonished when Halley's nucleus flared to 100 times its usual brightness. The first to spot this unprecedented outburst were astronomers Olivier Hainaut and Alain Smette, using the European Southern Observatory's 1.54m Danish telescope at La Silla in Chile. When the images were studied in detail, it could be seen that a cloud of dust 300,000km long ejected by the nucleus was reflecting sunlight and was responsible for the increased brightness. But astronomers were baffled by the cause of Halley's unexpected eruption. Models of cometary behaviour suggest that icy nuclei enter into a deep-frozen state when located much further than the orbit of Jupiter and are completely inactive at temperatures lower than -200°C.

Some astronomers believe that Halley's nucleus had collided with a large meteoroid (several metres across), and bits of Halley's crust at the impact site had been blown off to form the observed temporary cloud of material. Halley's nucleus may have been severely fractured during the impact and weakened so much that it might fail to maintain its integrity during its next visit to the inner solar system in 2061 and will break up before our descendants' eyes. Whatever happens, Halley will then be under the scrutiny of numerous sophisticated spaceprobes – perhaps even a manned mission dedicated to monitoring the activity of the comet's nucleus at close quarters.

There has been speculation that other comets have experienced temporary outbursts or even destruction as a result of impacts with meteoroidal debris. The splitting of Biela's Comet in 1842 (see page 19) may have been caused when it experienced a head-on collision with the Leonid meteor stream. According to research conducted by John Matese and Patrick Whitman of the University of Southwestern Louisiana, a number of flaring events observed in several 'fresh' Oort Cloud comets happened when they crossed known major meteor streams. For example, Comet Honda (1955) flared to magnitude 2.5 some 25 days following its closest approach to the Sun. Only a couple of days previously, the comet had ploughed into the Perseid meteor stream at a relative velocity of some 80km per second. The impact of one small meteoroid during this time may have carved out a small 1m-diameter crater in Honda's crust, which exposed volatiles inside the nucleus. Even a relatively small change on the outside may lead to greater changes in the long run, as the freshly exposed volatiles react to the Sun's energy and affect the local nuclear material. Weaknesses within the comet's crust may then allow buried pockets of gas to expand and cause a major outburst.

Comet Bradfield (1987 S)

Comet Bradfield was discovered by the veteran Australian comet observer Bill Bradfield in September 1987. For two months the comet passed through the northern constellations of Aquila to Pegasus, moving slowly northwards in declination. Bradfield was a delightful little comet to observe, with a well condensed head and a small, thin tail, and it proved to be brighter than had originally been predicted.

Comet P/Brorsen-Metcalf

This comet, which has now had three certified apparitions, was first observed in 1847 by Brorsen at the Altona Observatory in Hamburg, Germany. Its next return was in 1919, when it was spotted by Metcalf. The two astronomers, observing the same object but separated by 72 years, share the comet's name. In July 1989, the comet was recovered by Eleanor Helin using the 46cm Schmidt telescope at Mount Palomar.

From July to September, Brorsen-Metcalf carved a path across the sky from the northern constellations of Perseus through Auriga, Gemini, Cancer and Leo. My own observations of the comet were made in the pre-dawn skies of August, several weeks before perihelion, and at this stage the head of the comet was just visible to the naked eye. Using a 150mm rich-field reflector at a low power, the comet displayed a beautiful condensed nuclear region and sprouted a lovely 'spring onion' tail.

Comet Levy (1990)

In May 1990, using a 400mm reflector, American astronomer David Levy discovered this comet when it was a tiny 10th magnitude condensation in Andromeda. As the comet traversed the familiar asterism of the Square of Pegasus, it brightened, so that by June it had a small, rounded coma and a tail measuring 1 degree, looking much like a Spanish onion. By August, Comet Levy was easily visible with the naked eye, and it made an impressive sight through binoculars. Particularly noteworthy was the night of 18 August, when the comet passed just a few minutes of arc away from the globular cluster M15 in Pegasus. On 25 August, I observed Comet Levy move an angular distance of around 10 minutes of arc against the background star field, a real-time phenomenon of motion which I had only previously noticed with Comet Iras-Araki-Alcock, over seven years before.

Comet Shoemaker-Levy (1991 A1)

American astronomers Gene Shoemaker and David Levy discovered this comet in October 1991, when it was a tiny 16th magnitude dot on a photographic plate. From May to July 1992, the comet tracked through the northern skies from Cassiopeia to Ursa Major, but never brightened enough to be visible with the naked eye. Through binoculars and telescopes, Shoemaker-Levy appeared as a small, fan-shaped smudge of light with a tiny, star-like nuclear condensation.

Comet Swift-Tuttle (1992 T)

The parent comet of the Perseid meteor stream was recovered in September 1992 by the Japanese amateur astronomer Tsuruhiko Kiuchi, who used a pair of giant 25 x 100 binoculars to spot Swift-Tuttle, then a faint 11th magnitude object in Ursa Major.

It appears that earlier astronomical calculations based on a 120 year orbital period and predicting a return between 1979 and 1982 were inaccurate. Astronomers once identified Comet 1790 II (observed by Pierre Méchain) as the same object as Swift-Tuttle, and based some of their calculations on this erroneous assumption. But astronomer Brian Marsden reckoned that a comet observed by Kegler in 1737 was the true past manifestation of Swift-Tuttle, and on that basis he calculated an orbit which brought the comet back to our vicinity in 1992. Marsden was proved right, and incredibly one of his two predicted dates of perihelion passage (11 December) was just a day out. Such a success has prompted serious suggestions that Swift-Tuttle should be renamed Marsden's Comet, just as the work of Edmund Halley was honoured by naming the once-in-a-lifetime cometary visitor after him.

Seen from the Earth, Swift-Tuttle was never a very large comet, but it was easily visible with binoculars and went on to develop a short tail (around 1 degree in length if seen from light-polluted urban areas) as it coasted across the evening skies towards its closest encounter with the Sun at perihelion.

Comet Hyakutake (1996 B2)

In January 1996, Japanese amateur astronomer Yuji Hyakutake discovered the comet which bears his name using a pair of 25 x 150 'light bucket' binoculars. Fortunate viewers in the northern hemisphere followed the comet with great enthusiasm. Hyakutake rapidly reached naked-eye visibility as it sped through the constellations of Boötes, Ursa Minor and Cassiopeia during March, becoming its most brilliant in late March, when it had developed a large coma some 0.5 degree across and a tail which could be traced for 25 degrees from dark-sky sites. The comet and its tail were easily visible to the naked eye from light-polluted urban areas, and using binoculars the comet appeared to have a distinct greenish hue. High above the polluted atmosphere, the Hubble Space Telescope was swung into action and managed to obtain some exquisite views of the region around the nucleus, showing fine jets of dust and gas, as Hyakutake passed only 9.3 million km from the Earth. The comet was (then) the finest comet to have appeared since Comet West of 1976, and was the brightest one to pass so close to our planet in four centuries. On 26 June 1927, Comet Pons-Winnecke came closer than Hyakutake, at its nearest just 5.85 million km from the Earth. However, Pons-Winnecke's tiny inactive nucleus only attained a brightness of magnitude 13 and was of observational interest solely to professional astronomers and amateurs with large telescopes and dark skies.

To professional astronomers, Comet Hyakutake was noteworthy for its emission of copious quantities of methane and ethane gas – the first time the latter had been detected in a comet. Proportionately, the quantity of these substances was measured to be 1,000 times the amount generally thought to have been present in the early solar nebula. This apparent anomaly prompted speculation that Hyakutake had originated in another part of the galaxy, where the mix of raw comet-building materials was different than it had been in our own neighbourhood.

Adding further to the unusual nature of Hyakutake, the joint ESA/NASA Solar and Heliospheric Observatory (SOHO) – from its orbiting post at a gravitationally stable area called a 'Lagrange Point' some 1.5 million km from the Earth – observed tremendous quantities of water vapour pouring off Hyakutake's nucleus. The comet was losing no less than 3 tonnes of water vapour every second as the icy nucleus sublimated on its approach to the Sun – enough water to meet the demands of a medium-sized industrial town.

The Glorious Hale-Bopp

If astronomers thought they had been treated to an unbeatable once-in-a-lifetime spectacle with Hyakutake, they were in for a surprise when Comet Hale-Bopp began to advertise its celestial presence to all in the months following February 1997.

Comet Hale-Bopp was discovered independently on 23 July 1995 by American astronomers Alan Hale (a professional) and Thomas Bopp (an amateur), when it was a faint magnitude 10.5 patch of light in the middle of Sagittarius. It was soon found to be a very distant incoming object, well beyond the orbit of Jupiter at discovery. Further investigations showed that the comet had actually been imaged by the UK Schmidt Telescope at Siding Spring in Australia as far back as April 1993, when it was a mere 18th magnitude.

Being so bright and yet so far out at the time of its discovery, Hale-Bopp was allocated a brightness curve prediction of stunning proportions. Astronomers began to get excited at talk of a magnitude -1.7 coma (brighter than Sirius, the brightest star in the sky) at the time of its closest approach to the Earth. At the same time, they were cautious in accepting even the most pessimistic predictions because of past experiences with other comets that had failed to live up to expectations.

As Hale-Bopp came nearer, the experience of the infamous 'damp squib' Comet Kohoutek of 1973–4 was re-lived in the minds of many astronomers. Like Hale-Bopp, Kohoutek had also been discovered (by the Czech astronomer Lubos Kohoutek) as a very distant, bright-looking comet many months before it was due to arrive in the inner solar system. Astronomers had then been eager to accept optimistic predictions that Kohoutek would become the 'comet of the century'. In the end, expectations were dashed when the comet proved to be far dimmer than anticipated, reaching an observable maximum of just magnitude 6 with a 2 degree tail before perihelion, therefore becoming only just visible to the naked eye from really dark sites.

Thankfully, Hale-Bopp was different. By late 1996, the visitor had brightened enough to be seen through binoculars in the western evening skies after sunset. The comet was lost to sight for a short while as it became immersed in the glow of sunset, but by early 1997 it had cleared the Sun and for the first time had become visible to the naked eye, tail and all, as it passed through the constellation of Cygnus. Astronomers in the northern hemisphere could not believe their eyes as the true claimant to the title of 'comet of the century' brightened and established itself as a prominent circumpolar object. This time, the various news media were not raising false hopes when they encouraged people to go outside and spot the comet high in the western evening skies. From urban sites, the comet easily shone through the perpetual orange blanket of the streetlights' glow, and the brightest part of the tail, around 1 to 2 degrees long, was easy to see. From darker sites, the tail could be discerned as having two components – a curved yellowish dust tail approaching 10 degrees in length and a longer, straight, blue gas tail.

Telescopic views of Hale-Bopp's coma showed a bright, star-like nucleus surrounded by several distinct arcs of light – shells of dust and gas being swept back by the solar wind. The Hubble Space Telescope secured some fantastic close-up images of the jets arising from the 20km-diameter rotating nucleus.

Hale-Bopp's appearance in the closing years of the twentieth century provided the ideal opportunity for various astrologers, cranks, end-of-the-world merchants and an assortment of charlatans to claim that the comet represented a supernatural portent of tremendous change for the Earth, the affairs of humanity and the fate of important individuals. A group of computer software engineers and Internet technicians, calling themselves the 'Higher Source', arranged a bizarre mass suicide at their American mansion in March 1997. The strange cult had deluded themselves into believing that Comet Hale-Bopp's appearance was a sign for them to leave their earthly bodies and join a large alien spaceship travelling in the comet's wake. The deaths of the elderly Chinese communist leader Deng Xiaoping (born 1904), Pluto's discoverer Clyde Tombaugh (born 1906) and the famous scientist Carl Sagan (born 1935) all happened in early 1997 and were somehow linked with the comet. Likewise imagined to be the result of some mysterious celestial power, in the UK John Major's Conservative Party was ousted from office with the resounding victory of Tony Blair's Labour Party at the British General Election in May that year, when Hale-Bopp was still visible to the naked eye in the evening skies. It really is amazing what a large lump of dirty ice can do.

CLOSE ENCOUNTERS OF THE COMETARY KIND

There are three different ways in which the Earth can experience a close encounter with a comet. In the first scenario, the Earth can pass through the dust or gas tail of a comet whose nucleus might be many millions of kilometres closer to the Sun. At such an encounter, the comet would appear in the general direction of the Sun and therefore it, along with its tail, would not usually be visible around the time of the event.

This happened in May 1910, when the Earth made its way through the tail of Halley's Comet, but no ill-effects came about as a result of this global immersion. However, observers noted that the night sky seemed somewhat brighter than usual, as if it were covered with a thin veil of mist, and haloes were seen around the Moon and Sun. Enhanced rates of meteors were not observed, because the Earth ploughed largely through Halley's relatively dust-free gas tail.

Secondly, a comet's orbit may bring it particularly close to the Earth –

near enough for our planet to brush through the dust-rich coma surrounding the comet's nucleus. A major meteor storm would probably take place, as billions of cometary dust grains burned up in the Earth's atmosphere. In these kinds of close encounter cases, the comet's path would be altered drastically after its near-Earth passage, possibly resulting in a shorter orbital period which would compel the comet to remain within the inner regions of the solar system (like Encke's Comet, with its 3.28 year orbital period). Although such a set of circumstances has never been recorded, it is entirely possible, although not very likely. If the comet made its pass as it came from the outer reaches of the solar system, the whole event could be enjoyed (in the same way that some people enjoy a terrifying rollercoaster ride) as the nucleus and surrounding coma was observed to grow bigger night by night. At a distance of around 50,000km, an average-sized cometary nucleus would appear as an object about one-tenth the diameter of the Moon, surrounded by a coma which would take up much of the sky.

Collision Course

A collision with a comet's nucleus is the worst imaginable cometary encounter. Back in the 1950s, astronomers considered cometary nuclei to be small, flimsy and insubstantial objects – little more than 'flying sandbanks', which were incapable of doing much damage to larger bodies in the solar system. But we now know that even if cometary nuclei are in effect just large, dirty snowballs, their impacts can cause a great deal of havoc – enough to give the giant planet Jupiter a visible bruising.

A small cometary impact with the Earth would be akin to the above-ground explosion of a large nuclear device, and (apart from radiation) would produce many of the same devastating effects. Because three-quarters of the Earth is covered with water, the chances are that any future impact will happen over the ocean surface. This is likely to produce gigantic tidal waves, which would swamp coastal sites many thousands of kilometres from the impact site. A continental impact would have the potential to cause grave loss of life, not to mention global after-effects which might include climatic change of unknown duration or severity. It is believed that in June 1908 the impact of a tiny comet nucleus devastated a large but remote area of Siberia (see page 75). By remarkable luck this impact happened at one of the least-populated sites on the planet, and no great loss of human life was reported.

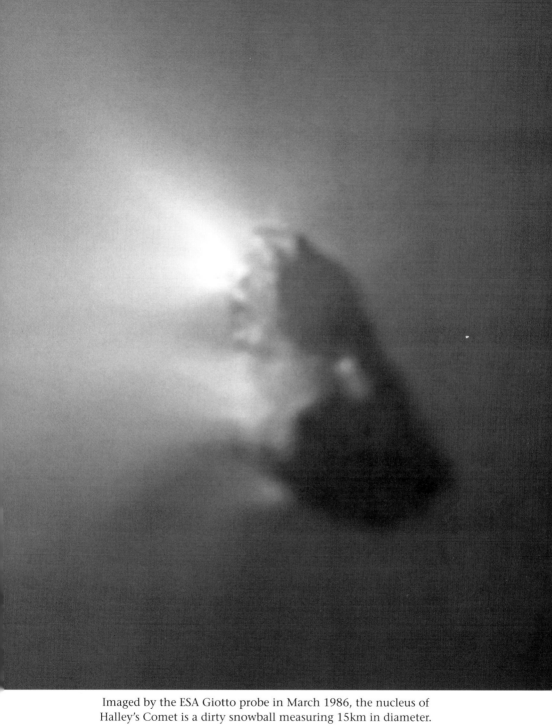

Imaged by the ESA Giotto probe in March 1986, the nucleus of
Halley's Comet is a dirty snowball measuring 15km in diameter.
European Space Agency

left The Leonid storm of 17 November 1966. Area shown is near the bowl of the 'Little Dipper'. *A. Scott Murrell, New Mexico State University Observatory*

above A woodcut depicting the Leonid storm of 13 November 1833. It was used to illustrate a nineteenth-century American biblical tract. *Courtesy Sky and Telescope Magazine*

below This advertisement of an angelic figure sprinkling mustard dust from space appeared in newspapers in October 1899 shortly before the predicted Leonid storm. *By kind permission of Unilever Historical Archives*

Comet Hale-Bopp, one of the brightest comets to appear
this century, made a spectacular showing in the spring
skies. Photographed 10 April 1997. *Peter Grego*

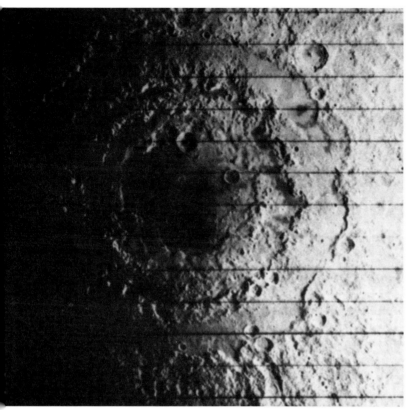

Mare Orientale is a huge lunar 'bullseye' – the scar of a mighty asteroidal impact that can occasionally be glimpsed by observers on Earth. This view was obtained by Lunar Orbiter 4 in May 1967. *NASA*

Impression of a large asteroidal impact on Earth. *Don Davis/NASA*

right The 31-tonne Ahnighito meteorite which was found in Greenland in 1894. The painting, based on an actual photograph, shows the meteorite being carted through the streets of New York to the American Museum of Natural History. *John Greco*

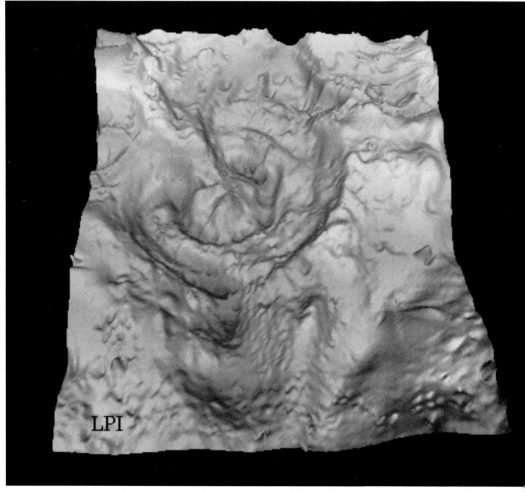

A computer model of the crater, lying beneath the Caribbean, believed by many to represent the scar of the asteroidal impact which wiped out the dinosaurs 65 million years ago. *NASA/LPI*

left Aerial photograph of the famous 49,000-year-old Barringer crater in Arizona, created by a meteoroid impact. It measures 1.2km across. *Dr David Roddy/US Geological Survey*

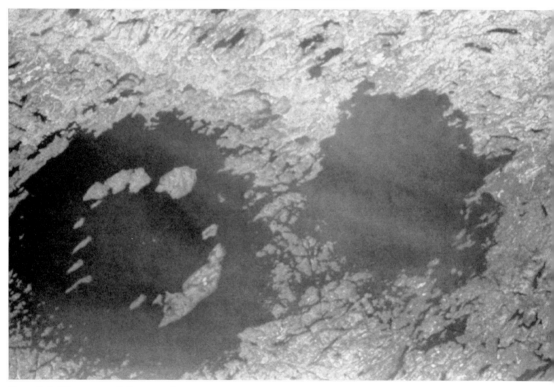

Canada's Clearwater Lakes are the remnants of a double meteoroid impact that took place around 300 million years ago. Clearwater West is the larger of the pair and measures 32km in diameter. *NASA/LPI*

Manicouagan crater, Canada. Measuring 70km in diameter, this water-filled ring is one of the Earth's largest well-preserved impact features and is about 200 million years old. In this view, taken from the shuttle Discovery in May 1991, the lake within the crater is frozen and appears brilliant white. *NASA*

Wolfe Creek Crater, Western Australia. This well-proportioned impact crater was discovered by oil prospectors in 1947. It measures 859m across with a rim that rises 50m above its floor. The feature is believed to have been blasted out of the desert rocks by a large meteoroid around 300,000 years ago. *Courtesy Virgil L. Sharpton*

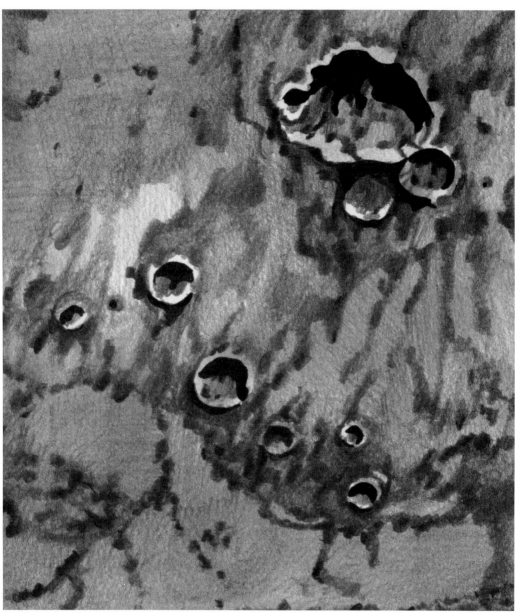

Drawing based on an aerial photograph of the Henbury craters. 130km southwest of Alice Springs, this cluster (known locally as the Devil's Punchbowl) nestles in the Australian outback, evidence of the prehistoric impact of a large iron meteoroid which broke up in the Earth's atmosphere. *Peter Grego*

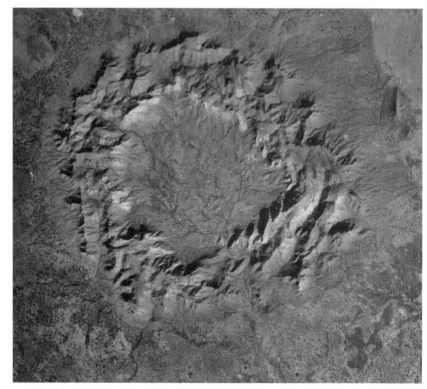

above Aerial view of the Gosse's Bluff impact feature in Australia's Northern Territory. The ring of mountains, which measures a total of 22km in diameter, is actually the exposed remnants of an asteroidal impact that took place around 140 million years ago. *Commonwealth of Australia, AUSLIG, Australia's national mapping agency. All rights reserved. Department of Administrative Services, Canberra ACT*

below Asteroid Mathilde, measuring 52km in diameter. Image taken by the NEAR probe in June 1997. *NASA*

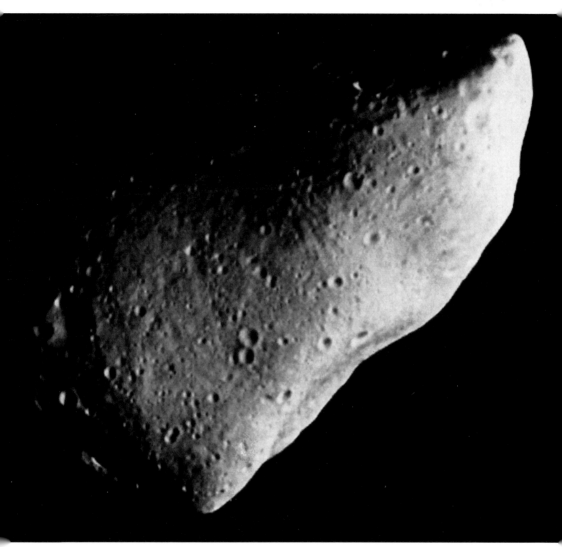

Asteroid Gaspra, measuring 18km in length. Image taken
by the Galileo probe in June 1992. *NASA*

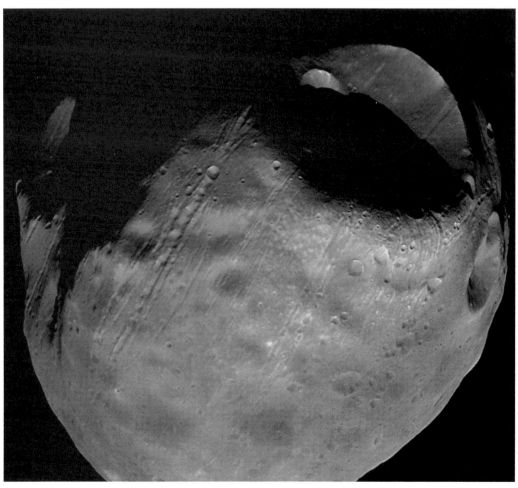

The innermost satellite of Mars is a captured asteroid. Its orbit is slowly deteriorating, and in about 30 million years time Phobos will crash into the Martian desert. *NASA*

Impression of Comet Shoemaker-Levy 9's impact
with Jupiter in July 1994. *Peter Grego*

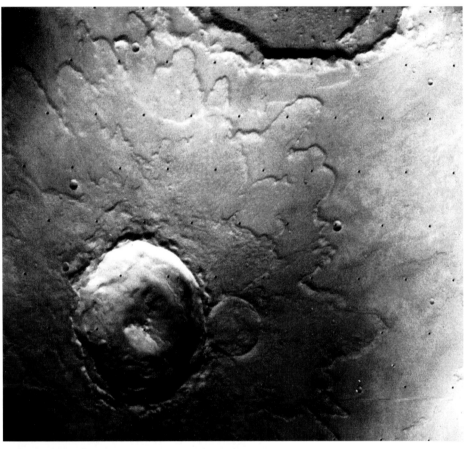

above View from the Viking Orbiter, 1976, of the 'mud-splash' pattern of ejecta on Martian crater Yuty. *NASA*

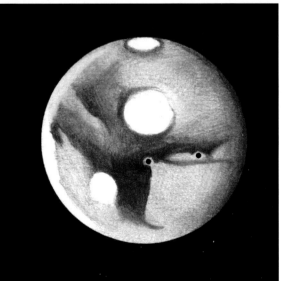

left Observation made on 30 October 1988 using 300mm reflector. The image shows three Martian impact craters – the vast Hellas basin (the large bright circle upper centre of the disc) and the craters Huygens and Edom which appear as small notches in the dusky desert markings (positions indicated).
Peter Grego

First photograph of the residual atmospheric effects caused by a fireball that streaked across South African skies in 1912, taken from Union Observatory Circular. *Courtesy BAS Library*

Author's observation of spectacular fireball as it appeared in the skies on 17 April 1984. *Peter Grego*

Image taken by the POLAR satellite, 26 September 1996, showing the trail left by a microcomet, 12m in diameter, as it vapourized in the Earth's atmosphere at a height of over 1,000km above the Atlantic Ocean and Western Europe. *NASA*

2

METEORS: LIGHTS IN THE SKY

INTERPLANETARY DEBRIS

Often, the interplanetary space surrounding the Sun is conveniently portrayed as a void containing nine major planets and a few assorted moons, asteroids and comets. But that is rather like describing the room around you in terms of what you can readily see. Unless you happen to be in a laboratory, operating theatre or spaceprobe clean-room facility, on looking closer you will find quantities of dust, even in the cleanest-looking places; turn the light off, shine a flashlight around and you'll see that the air is full of floating particles. There are bound to be a one or two insects visible to the naked eye lurking around, and with a powerful microscope it would be possible to discover countless other life-forms.

The same is true of interplanetary space – there's more material hanging around our local planetary neighbourhood than we care to imagine – plenty of cosmic dust, meteoroids and assorted debris of one kind or another, some of it man-made. The life-forms are restricted to a few select places, mainly on the Earth, but spots of terrestrial bacteria and other hardy micro-organisms also cling precariously to various unsterilized bits of rockets and spacecraft on the Moon, Mars and in space. Astronomers reckon there is a chance that primitive life may also have developed independently on Mars, inside the subcrustal water oceans of Jupiter's moon Europa, and perhaps within the rich gaseous atmosphere of Jupiter itself.

There are several ways in which meteoroids and interplanetary dust can be detected. In the above analogy the flashlight can be seen as the light of the Sun. Its rays illuminate quantities of material in interplanetary space, and tracts of dust spread along the plane of the ecliptic (the plane along which the planets orbit) can be seen as a faintly glowing nebulosity. This is evident in the Zodiacal Light, which can be seen from dark-sky sites as a pearly cone of light lying along the ecliptic, rising from the western horizon after sunset in spring or before sunrise in the eastern morning skies of autumn.

Another related phenomenon, the gegenschein (from the German for 'opposite shine'), can be seen as a faint nebulosity diametrically opposite the sun on the ecliptic and is therefore best placed at midnight in exceptionally dark late-winter skies, when it is at its highest point above the horizon. Both the Zodiacal Light and the gegenschein are produced by sunlight's reflection off particles of dust 0.2–0.1μ (micron)/0.0002–0.0001mm in diameter, and the sight of these phenomena forcibly makes the point that interplanetary space is by no means empty.

Meteoroidal Bombardment

It has been estimated that the Earth's atmosphere is bombarded by up to 200,000 tonnes of meteoroidal material every year, but most of this is completely vaporized by the intense frictional heating of the Earth's atmosphere. Only the smallest of microscopic particles, called micrometeoroids, succeed in encountering the Earth's atmosphere from the depths of space without being subjected to severe frictional heating, and some 30,000 tonnes of this microscopic interplanetary dust is accumulated every year, most of it landing on the oceans and sinking to the sea bed. Each year, just 2,000 tonnes of larger meteoroidal particles survive the heat of re-entry and land on the Earth; these solid, tangible objects from space are known as meteorites. At this rate of accumulation (over 5 tonnes per day), spread evenly over the Earth, it would take a billion years to build up a 1mm depth of material!

A meteor – commonly referred to as a shooting star – is the name given to the flash of light observed in the sky when a meteoroid plummets towards the ground and frictional heating by the Earth's atmosphere causes it to glow brightly and rapidly erode, heating the air around it and leaving behind a shining dust and gas trail. Most meteors are visible to the naked eye when they burn up at heights between 200km and 50km above the ground. The faster the meteoroid, the higher in the atmosphere it will burn up, since it is hitting more atmospheric molecules per second than its slower counterpart. Leonids, the fastest of the annual showers, travel at a phenomenal 71km per second and burn up in the Earth's atmosphere between 120km and 100km above the ground. The slowest of the shower meteors can penetrate much deeper into the atmospheric shell before burning up, typically appearing between 85km and 70km high.

Meteors shine at their brightest during their brief lives when about 70 per cent of the incoming meteoroid has been burned away; the level of actual brightness depends on the mass and velocity of the meteoroid (its composition plays a minor role in brightness) and the angle at which it approaches the Earth's atmosphere. The apparent brightness varies, of course, with the distance of the observer, the meteor's elevation above the observer's horizon and the conditions of the atmosphere. It has been calculated that a meteoroidal mass of 1g, travelling at a velocity of 30km per second (108,000km per hour) and entering the Earth's atmosphere at an angle of 45 degrees, will produce a zero magnitude meteor to an observer located at a distance of 100km.

If you stand outside on any clear night, on any day of the year, and look upwards in any direction for around half an hour, the chances are that you will see at least one meteor streak across the sky. Of course, your chances of success will improve dramatically if you are located at a dark site well away from streetlights and other light pollution – including the Moon, an object whose light is often cursed by meteor observers. Some of these meteors may be classified as 'sporadics' – solitary particles of cosmic dust which happen to burn up in the Earth's atmosphere at unpredictable times. A staggering total of some 90 million (potential) naked-eye meteors brighter than the fifth magnitude appear in the Earth's atmosphere every day.

On rare occasions, the Earth encounters previously unknown clusters of meteoric debris which produce unique, unrepeatable star-storms. This happened on 5 December 1956, when an unexpected downfall of meteors, apparently emanating from the constellation of Phoenix, was observed in the skies of New Zealand, Australia and South Africa. Stunned watchers reported hourly rates of up to 100 meteors, many of which were exceptionally bright fireballs – as bright as Venus – leaving conspicuous glowing trains in their wake.

Aside from such unpredictable, isolated, sporadic meteors and rare showers, every year the Earth encounters dozens of distinct streams of meteoritic debris which have been deposited in space by comets. These streams produce annual meteor showers that take place at around the same dates year after year. There are currently more than a dozen annual meteor showers which can reliably be observed from the northern hemisphere.

It would be quite wrong to imagine that comet dust burning in the atmosphere would appear monotonously identical time after time. No two showers are alike, nor are their individual meteoric displays. Some showers produce fast, brilliant meteors, while others give rise to slow-moving meteors which leave persistent glowing trails in the sky behind them. Meteor colours are commonly reported, due as much to the way in which the human eye perceives flashes of light in the night sky as to the speed and substance of the meteoroid and its effect on the atmosphere.

A good meteor shower can be imagined from another perspective. Go to the centre of your sitting room, hold a cushion above your head and beat it vigorously. Turn off the main lights and shine a torch in front of you. Now walk across the room very slowly (literally at a snail's pace), and imagine yourself to be the Earth, 12,700km in diameter from head to toe, encountering a dense shoal of meteoroids whose individual members are separated from each other by about 10km. Of course, the house-dust meteors would need to be moving at a uniform pace in one direction and parallel to each other in order for this analogy to be perfectly accurate, but the concentration of particles in relation to your size is about right for a splendid display for the two observing stations (your eyes) located at temperate facial latitudes.

Meteors can be exceptionally brilliant. Above magnitude -4 (as bright as the planet Venus at its most spectacular showing), meteors are classed as fireballs or bolides. They are produced when larger meteorites (bigger than a grain of rice) enter the atmosphere; their outer surfaces melt and ablate, leaving behind streams of glowing particles in their wake. The head of the fireball can be dazzling, and much of this intense brilliance is produced when the compressed envelope of air in front of the meteoroid becomes superheated and luminous, attaining a diameter thousands of times larger than the meteoroid itself.

ANNUAL EVENTS

A few annual showers are renowned for the numbers of bright meteors and fireballs they produce, some with persistent trains which hang around in the atmosphere for a while. Each shower has its own particular delights, and every one of them is quite capable of springing surprises from year to year. Most of the annual meteor showers listed below make their way (moonlight permitting) on to the observing list of every die-hard meteor observer in the northern hemisphere.

The Zenithal Hourly Rate (ZHR) is a theoretical value given for the number of meteors a single observer would see if the radiant were directly overhead in a dark-sky site with a limiting magnitude of 6.5 (on the very limit of naked-eye visibility). Naturally, this is a set of circumstances rarely granted in combination, meaning that actual observed rates are usually much lower than the ZHR.

Quadrantids

The honour of being the year's first meteor shower goes to the Quadrantids, which make their appearance on the very first day of each new year. The meteors take their name from the long-defunct constellation of Quadrans Muralis, from which they were once seen to radiate.

Quadrans Muralis is the Latinized name for an obsolete piece of astronomical equipment known as a Wall Quadrant. The constellation was first introduced into the sky by the German astronomer Johann Bode (1747–1826), who included this quaint grouping in his star atlas

Uranographia (1801); the faint stars of Quadrans lie within, and to the northeast of, the constellation of Boötes the Herdsman, past the end of the Great Bear's tail on the border of Draco and Hercules. Alas, Quadrans was a rather unspectacular constellation containing nothing of importance, not even a bright star, and it is difficult to locate even with the aid of a star map. It isn't surprising that later generations of star-watchers chose to ignore Bode's unusual stellar creation in favour of the more ancient and traditional groupings in this particular region of the heavens. There have been numerous attempts to rename the shower the Boötids, but the original nomenclature stubbornly remains, a reminder of times when the naming of celestial objects – romantic though many of the old names were – was somewhat arbitrary and unregulated.

Quadrantid meteors, many of which are faint, shoot across the sky at a medium speed, and many have been noted to possess a bluish tint. About one in ten Quadrantids leaves behind a glowing train. The shower takes place from 1 to 5 January, with maximum occurring on 4 January, and a ZHR sometimes exceeding 100. In 1965, a Quadrantid rate of 190 was observed, and in 1973 and 1985 there were ideal conditions for the Quadrantids when the Earth passed through a dense part of the meteor stream, during which time high rates were seen. Undoubtedly the finest recorded Quadrantid display took place in 1909, when it achieved an impressive rate of over 200 meteors per hour.

The Quadrantids have been seen in the Earth's skies since the early nineteenth century, and although they are sometimes claimed to have first been recognized as an annual shower by Louis François Wartmann in Geneva, Switzerland, in 1841, the Quadrantid stream was actually discovered during the previous decade by Lambert Quetelet, mathematician, statistician and one-time Director of the Brussels Observatory in Belgium, who was responsible for naming the shower. Certainly, before the nineteenth century the Quadrantid stream did not cross the Earth's orbit, but gravitational perturbations induced by other planets eventually shifted the stream into our path.

Further perturbations will ultimately remove the stream from its intersection with the Earth's orbit, and it is thought unlikely that any Quadrantid meteors will be seen after the twenty-second century has got underway. It is considered improbable that any more spectacular displays will be put on by this stream, and the 1909 event probably represented the Quadrantids' finest moment.

No Quadrantid parent body has been identified, although Comet Macholz, discovered in 1986, is a possible candidate. Alternatively, according to the work of Iwan Williams and Zidian Wu of Queen Mary and Westfield Colleges in London, a comet observed in 1491 by Chinese astronomers possessed identical orbital elements (as far as can reasonably be deduced) to the Quadrantid stream, and may be the same comet as one seen in 1385. After a close encounter with Jupiter in the middle of the seventeenth century, the comet may have been perturbed into an altogether different orbit so that it no longer approaches Earth.

Some research suggests that the original cometary parent of the Quadrantids disintegrated in two episodes around AD 300 and AD 700, creating the meteor stream we see today. There may be large chunks of the parent body still remaining in the solar system which are inactive and have not yet been identified; the chances of their hitting the Earth (if indeed they do exist) are slim, but such a catastrophe cannot be ruled out entirely.

April Showers

The Virginid complex, a group of related meteor showers, produces rather low rates from February to mid-May. These meteors, radiating from around the constellation of Virgo, are very under-observed because they are infrequent and spread over many weeks, and accurate details regarding this shower group are unknown. The best known Virginid substream has a radiant near Spica (Alpha Virginis) and is active between 7 and 18 April, peaking on 12 April with a ZHR of about 5. The Virginids are slow meteors, occasionally display flaring, and some 12 per cent of them leave persistent trains.

By far the most noteworthy of the spring meteors, the Lyrid shower from 19 to 26 April produces medium-speed, brilliant meteors, a quarter of which leave glowing trains; many fireballs may be seen during a good Lyrid display. The shower has a fairly reliable ZHR of about 15, which peaks on 20 to 21 April; the radiant climbs to a respectable altitude during the night and is highest by dawn.

The Lyrid stream is a mass of debris left by Comet Thatcher of 1861, a comet with a 415-year period. There are records of Lyrid storms in the Far East in 687 BC and 15 BC. The last true Lyrid storm was seen from the eastern US in 1803, where a rate of 700 meteors per hour was logged during a two-hour burst of activity. In 1922, enhanced activity was observed from Greece, during which rates of approximately 100 per hour occurred, and at Florida in 1982 a short burst in excess of 80 per hour was recorded. The Lyrids are certainly well worth keeping a watch for, since the stream may contain large concentrations of material strung along its orbit at irregular intervals, which are making their unseen way towards inevit-able encounters with the Earth's atmosphere – more spells of enhanced activity or even a major storm are not beyond the realms of possibility.

At the date of Lyrid maximum, another meteor shower begins to produce activity. The Eta Aquarids, very fast meteors with persistent trains, begin around 21 April and last until 25 May, during which time the radiant moves from the border of Capricornus, through Aquarius at maximum and into Pisces. The maximum is broad and straddles either side of 3 May, giving rates of up to 40 per hour. However, the shower is not well placed for British observers because the radiant is very low during the hours of darkness, and further north the situation worsens – from Scotland the light of dawn is already spreading in the east by the time the radiant manages to rise above the horizon.

An eruption of the nucleus of Halley's Comet around 4,000 years ago is said to have been responsible for depositing the material which makes up the Eta Aquarid stream. This debris is encountered by the Earth as the stream material is moving along the outward-bound track of its orbit. It is ironic to realize that just one modest speck of burning meteor dust from Halley can appear brighter for one brief moment than the entire comet at its best during the 1985–6 apparition!

The first recorded Eta Aquarid shower took place in 74 BC, but owing to the unfavourable position of northern hemisphere observers, the shower was not very well studied until southern hemisphere meteor observers took up the challenge and began to organize serious watches in the 1920s. Even before this, it had been established that the Eta Aquarids are part of the same meteor stream as the Orionids (see page 43) which can be seen in October, and it is the only major stream encountered by the Earth twice in a year. The Earth passes closer to the more densely populated centre of the stream in May than it does in November (10 million km and 23 million km respectively), and therefore Eta Aquarid rates are higher than Orionid. Note that Halley's Comet no longer inhabits the same path as the debris it ejected around 2,000 BC, owing largely to numerous gravitational perturbations, combined to some extent with its own nuclear eruptions which have acted like rocket engines to alter its path.

Some astronomy texts mention a shower known as the Draconids, or 'Pons-Winneckids', which are said to be active from 20 May to 10 July, with a maximum around 30 June. Anyone who has recently spent an evening gazing in the direction of Draco hoping to spot a Draconid meteor on the basis of this information will have been acutely disappointed, because this shower has long been extinct! This Draconid stream, believed to have been deposited by Comet Pons-Winnecke (see page 29), was deflected from the Earth's orbital path many years ago, and the meteoroids of this stream no longer burn up in our atmosphere.

From May to July a little-known shower complex called the Ophiuchids produces low rates, peaking on 9 and 19 June with a ZHR of around 5. The meteors are slow moving and unspectacular, with long paths across the sky. Because this stream lies along the plane of the ecliptic – the plane in which the planets orbit – it is prone to frequent gravitational perturbations. These disruptions have caused the Ophiuchid streams to lose much of their original integrity. Around the same time there is some meteor activity from the Scorpiid-Sagittariid streams, which is best seen from the southern hemisphere, and they may share links with the Ophiuchid streams.

Summertime Meteors

Anyone who spends a while looking up at the sky on a clear summer's night will see the flash of a meteor cross the heavens. Summer in the northern hemisphere is a good season for meteor-hunting, since between early June and mid-August no less than ten meteor showers compete for attention – more if you count the Beta Taurids, a daytime shower detectable only by radio equipment, and the showers of the Ophiuchids and Piscis Australids visible from southern latitudes. This natural abundance of summer meteors is great for introducing people to astronomy in a painless way – shirt-sleeve observing conditions, no optical aid required and (almost) guaranteed success. However, the ardent amateur meteor watcher can sometimes feel overwhelmed by this seasonal overkill – many tend to concentrate on keeping a watch for a select few meteor showers, allocating their time and energy sensibly rather than attempting to take on the lot.

Perhaps some meteor observers secretly thank the Moon for providing a regular monthly break in their observing programme at times when it is bright and near-full, hence destroying the dark conditions required for meteor observation. The Moon's bright light can seriously affect meteor counts; a full Moon will allow only the brightest meteors to be spotted, and meteor observers always consult their astronomical diaries to see whether the Moon will be prominent at the dates of the various annual meteor showers. But should moonlight interfere with the date of a shower maximum, our satellite is more than likely to be subjected to some form of verbal abuse. The same nocturnal curses aimed at Luna from the lips of ardent meteor watchers are also uttered by deep sky observers. Some years ago I actually heard one amateur astronomer say 'I hate the Moon! I really detest it! You can't do any serious astronomy when that thing's around!' Such 'Mooncussers' may take some consolation from the fact that meteorites are the main cause of erosion on the lunar surface – needless to say, any hopes that the Moon will be rapidly scoured away to dust by impact are rather naïve!

With their radiant above Deneb (Alpha Cygni, one of the bright stars which make up the 'summer triangle' asterism), the Alpha Cygnids are one of six meteor showers which take place between July and August each year. The radiant is very high, almost overhead during the night, and the Alpha Cygnids are active between 12 July and 21 August, with a maximum ZHR of 5 on 5 August. However, less than one in ten achieves brilliance and leaves a persistent train in its wake.

Three maxima are displayed by the Capricornid shower – 8, 15 and 26 July – each with a peak ZHR of around 5. Capricornids, some of which can be brilliant, move very slowly across the sky and a small percentage leave persistent trains.

The Iota Aquarids are faint and slow meteors, active between 15 July and 25 August, and reaching their maximum on 6 August with a ZHR of 8. This shower has a double radiant, one north and one south, separated by about 10 degrees. Over the course of the shower period, the radiants move across the sky from west to east through Capricornus and into Aquarius.

To confuse meteor observers further, the Delta Aquarid shower takes place between 15 July and 20 August, and has two maxima – one on 29 July and the other on 7 August – giving ZHRs of about 20 and 10 respectively. Like the Iota Aquarids, the Delta Aquarids have a dual east-migrating radiant which lies to the north and south of the Iota radiants, separated by more than 15 degrees. Although the southern Delta Aquarid radiant is slightly more active than the northern, its lower altitude seen from Britain means that the observed rates here are about the same for both components. Most Delta Aquarids are faint, although there are sometimes brilliant examples; they travel in long paths at a medium speed and about 8 per cent of them leave persistent trains. Of particular note are the superb Delta Aquarid showers which took place in AD 714 and AD 784 – nothing to match these outbursts has been recorded since.

Not to be confused with the earlier Capricornid shower, the Alpha Capricornids show between 15 July and 25 August and reach a peak ZHR of 5 on 2 August. They are slow, distinctly yellow meteors, which can sometimes appear brilliant – a fact which recompenses the observer in northern latitudes, who is unlikely to see most of the shower's fainter examples. The Alpha Capricornid radiant starts in Sagittarius, traverses northern parts of Capricornus and ends in the stars of Aquarius.

Of all the summer showers, the Perseid shower is the best one observed from the northern hemisphere. Activity begins on 23 July and ends on 20 August; maximum occurs on 12 August with a peak ZHR of 75, and good rates can be expected a day or more on either side of this date. Through its period of activity, the Perseid radiant moves from its initial appearance just beneath the 'W' of Cassiopeia, through the double cluster of h and Chi Persei and terminates in Camelopardus. Perseid meteors are very fast and bright, often flare in mid-flight and leave behind lovely glowing trains. The Perseids are fairly reliable from year to year, but there was cause for concern in 1911 and 1912, when exceptionally poor rates led many to speculate that the shower was nearing the end of its life. Fortunately, reports of the Perseids' death were greatly exaggerated, for a decade later excellent displays gave rates as high as 200 per hour. Subsequent rates have been high, and notable showers took place in 1980 and 1981, with rates of 120 and 100 meteors per hour respectively. Near-storm conditions prevailed in 1991, where 300 to 400 meteors per hour were recorded in eastern Asia, and in 1992 hundreds per hour were observed (up to 400 per hour from eastern Europe) despite the interference of a bright gibbous Moon. Perseid rates observed from the Mediterranean during 1993 briefly reached many hundreds per hour, with a very high proportion of brilliant meteors and fireballs. To European and eastern US observers, the following year's Perseid display attained average rates; but in the western US, around 9:30 UT, very high rates of up to an estimated 250 meteors per hour were seen briefly.

In 1866, the Italian astronomer Giovanni Schiaparelli proved mathematically that the members of the Perseid meteor stream moved in the same path through space as Comet Swift-Tuttle (1862 III), and since then many other meteor showers have had their parent bodies deduced in the same manner. The Perseid stream derived from Swift-Tuttle is very old and is thought perhaps to have been around for more than 2,000 years. The earliest record of the shower comes from China in AD 36, and the first European account was given in AD 811. The Perseid shower was once commonly known as the 'burning tears of St Lawrence' because it peaked on 10 August, around the feast day of the Christian martyr who was executed on a grid-iron during the reign of Roman Emperor Valerian in AD 258.

Enhanced Perseid numbers were observed before the time of Comet Swift-Tuttle's expected return in 1982, but the comet did not make its appearance until a decade later, in 1992, proving that most of the old estimates of a 120 year period for the comet were incorrect. It was fascinating that year to observe the Perseid meteors around the time of their maximum on the night of 12/13 August, and less than three months later to conduct a series of observations of the shower's parent comet as it brightened to make its first appearance to the naked eye since 1862. It was a delightful little comet to observe with an optical aid, and it developed a nice tail, easily visible with binoculars, as it sped through the evening skies towards perihelion.

Slow, brilliant, flaring, yellow-coloured meteors are the trademark of the Kappa Cygnid shower which is active between 19 and 22 August; the shower reaches its maximum, radiant stationary and directly overhead, on the evening of 20 August with a ZHR of 5. One in three Kappa Cygnids shows a distinct train. On 22 August 1988 at 23:00 UT, from the coast of North Wales I observed a fine example of a brilliant, golden-coloured Kappa Cygnid meteor which flared to about magnitude -3, a splendid object which made its way serenely through the heavens and cast its light over the waters of Cardigan Bay.

Fast, bright meteors, similar in appearance to the Perseids, the Alpha Aurigids display themselves between August and October, with maxima on 28 August and 14 September. These meteors have a double radiant just to the right of the triangular 'kids' asterism in Auriga. Because the shower activity comprises mainly bright meteors and spans such a broad period of time, it is thought that the Alpha Aurigids are very old and are now slowly fading, but a ZHR of about 10 can still reasonably be expected. The Alpha Aurigids are best observed when the radiant is high in the morning skies before dawn.

Autumn's Falling Stars

One of the twentieth century's most spectacular meteor displays lasted only four hours and took place in European skies on the evening of 9 October 1933. During the period of maximum activity, observers at Potsdam in Germany counted 5,700 meteors in 30 minutes; a group of four observers at Hamburg-Bergedorf counted 345 in one minute, and a gentleman named Forbes-Bentley in Malta reported 480 in one minute. The brightness range of this splendid flourish of meteors went from that of Venus (magnitude -3) down to that of the limit of visibility. By the time the Earth had turned so that the radiant had been brought above the horizon for observers in the US, the period of frenetic activity had ended. The Draconid meteor shower (also called the Giacobinids after their parent, Comet Giacobini-Zinner, with a period of 6.6 years) is still active between 10 and 20 October, with a maximum on the first day of the shower. The rates vary considerably, with occasional rich displays; they are worth watching because of the possibility of another Draconid spectacular like that which astonished European eyes in 1933.

The stuff which makes up the Draconid stream was directly sampled when, on 11 September 1985, the joint NASA/ESA spaceprobe named the International Cometary Explorer (ICE) flew into the coma of Comet Giacobini-Zinner and came within 8,000km of its icy nucleus. The probe, originally launched in 1978 to observe the solar wind and named the International Sun-Earth Explorer (ISEE), had no dust shields like those of the specially designed European Giotto Halley probe. As it plunged into the comet's coma, small meteoritic impacts hit the probe at a rate of one per second, proving the coma to be far less dusty than expected. ICE took 20 minutes to cross the comet's 24,000km-wide tail and emerged more or less intact and in good working order. ICE's measurements of dust, molecules of water ice and carbon dioxide confirmed the theory that comets are large, dirty snowballs. The probe even went on to make observations of Halley's Comet from a more respectable distance in March 1986 – if only more spaceprobes would commit themselves to a life of such gallant and uncomplaining service!

In mid-October 1985, while patiently searching for Halley's Comet during those cold mornings, my attention at the eyepiece was distracted now and again by the flash of a bright meteor. I soon made the connection which still astonishes me to this day – those brilliant meteors were members of the Orionid shower, nothing less than tiny fragments deposited over the centuries by Halley's Comet itself.

Orionids are visible between 16 and 26 October, with a maximum on 21 October. The meteors, accelerating towards the Sun on the inward-bound section of their orbit, are bright and very fast, travelling at 66km per second; more than 30 per cent of them leave persistent trains. Peak Orionid ZHR is around 25, and fireballs are a common sight about three days after maximum. The Orionid stream is complicated, with fluctuations of activity from year to year; it has been found that the radiant is multiple, but because they are concentrated in a small area of sky this is not obvious even to the regular visual observer.

Ocurring between 20 October and 30 November, the Taurid shower produces slow meteors – at a speed of just 28km per second, this is the slowest of the major streams – and the shower is noted for its brilliant meteors and occasional fireballs. Peak activity takes place from 1 to 3 November, giving rates of up to 15 meteors per hour. The Taurids consist of a number of related meteor streams and, according to one theory advanced in 1950 by comet experts Fred Whipple and S. El Din Hamid, the stream originates from Comet Encke when it flared about 4,700 years ago, and is augmented by another ejection of material 1,500 years ago. The flares may have been produced by sudden activity in the icy nucleus, or (more likely) as a result of meteoroidal collision. There are two Taurid radiants: the southern radiant which runs west–east along a line passing through the Aldebaran side of the 'V' shape of the Hyades group, and the northern radiant which runs parallel to it and passes just underneath the Pleiades open star cluster.

The Leonid meteor stream lives along the path of Comet Tempel-Tuttle in an orbit inclined 17° to the Earth's orbital plane. The comet and meteoroids travel in a retrograde orbit, and each year Earth intersects this debris head on, with the observed meteors travelling at a velocity of 71km/second.

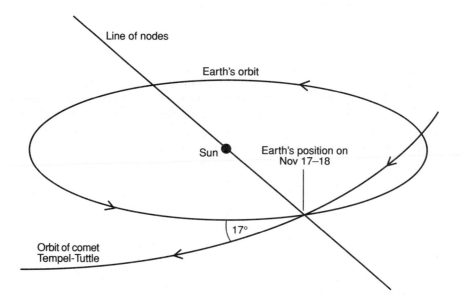

November skies over Britain are host to millions of firework rockets launched to celebrate the foiling of Guy Fawkes' 'Gunpowder Plot', a fiendish plan to blow up King James I (James VI of Scotland) and his parliament in 1605. As the man-made fireworks fizzle out, a display of interplanetary pyrotechnics called the Leonid meteor shower puts on its annual show. In summary, the Leonid shower produces very fast meteors which have persistent trains. Activity occurs between 15 and 20 November from a slow-moving radiant in the 'sickle' of the head of Leo, with a peak on 17 November. Normal Leonid rates reach about 15 per hour, and they are the fastest of the annual showers, with a speed of 71km per second. Comet Tempel-Tuttle (1866 I) is the known parent body; the comet rounds the Sun every 33 years and deposits large amounts of debris in space during the periods when it is in the Sun's vicinity. Every 33 years, when the Earth is near Tempel-Tuttle's nucleus, enhanced Leonid activity is observed as we plough into this debris; over a millennium of recorded Leonid activity has produced many spectacular Leonid showers and some incredible Leonid star-storms. (For a full account, see Chapter 3.)

Christmas Lights

Two very different meteor showers occur before Christmas – one which is currently the best of the annual showers, followed by another which produces very low rates but is noteworthy because it is unpredictable. The Geminids show between 7 and 15 December, maximum taking place a couple of days either side of 13 December, with a peak ZHR of 75. Geminids are intensely white meteors, very bright and slow moving. The radiant moves towards Castor (Alpha Geminorum) as maximum is approached. Geminid meteors have only been recorded for certain since 1862, with a constant rise in activity to the current high level. Using data obtained in 1983 from IRAS, astronomers Simon Green and John Davies discovered a new asteroid which was later named 3200 Phaethon. The asteroid's orbit closely matches the Geminid stream, and it is likely to represent the dark-crusted, burned-out nucleus of a once active comet which gave birth to the Geminids.

The Ursids, the year's last annual shower, are active from 17 to 24 December, maximum occurring on 22 December. The shower's radiant is located in Ursa Minor, and as it is circumpolar it is visible all night long. The Ursids were discovered in the nineteenth century by the British astronomer William Denning, but came into focus in 1945 when Czech astronomers (including A. Becvar) observed a burst of high rates exceeding a ZHR of 50. Since then, the Ursids have maintained a rate of about 10 to 15 per hour during a relatively short period on the night of maximum. In 1979, a further flourish of Ursids produced rates of 25 meteors per hour. Comet Tuttle of 1858 (first seen by Méchain in 1790) is thought to be the Ursid's cometary progenitor.

OBSERVING METEORS

Colours

Research carried out by Alastair McBeath, Director of the Meteor Section of Britain's Society for Popular Astronomy (SPA), has found little evidence to support the view that the colours displayed in sporadic (non-shower) meteors provide an insight into the nature of the meteoroids that produced them. The work, based on a total of 5,660 reliable meteor observations made by SPA members from 1984 to 1988, revealed that only 16 per cent of the total displayed distinct coloration. By far the biggest proportion of coloured sporadics were yellow, amounting to an impressive 80 per cent of the coloured total. McBeath attributes this, in part, to the poor yellow-white colour distinction of the human eye, where plain white meteors of medium magnitude have been mistakenly assigned a yellow colouration. There were fewer green-coloured sporadics than any other colour, with just six reported (0.7 per cent of the coloured total). Green meteors may result from meteoroids with a high magnesium (Mg I) content; spectra of very fast meteoroids have also shown the 'forbidden' green oxygen line, thought to be atmospheric in origin and produced by ionization, rather than revealing anything of the meteoroids' composition. McBeath concludes that the colour balance observed in the minority of sporadic meteors which do show colour is due to a number of possible effects, and the most significant of these appear to result from colour perception and contrast in the human eye.

A similar analysis of the Delta Aquarid, Perseid, Taurid and Geminid meteor showers, made by SPA Meteor Section members between 1984 and 1990, arrives at the same conclusion – that a high proportion of observed meteor colours is largely explicable in terms of the sensitivities and failings of the human eye. However, the Taurid shower seems to be genuinely rich in yellow meteors, totalling nearly 29 per cent of all observed Taurid meteors. The Perseids and Geminids had the highest proportion of blue meteors, this colour showing in more than one of every 20 meteors. Green- and violet-coloured meteors were the rarest type. The Geminids displayed the highest proportion of green meteors, with one for every 333 Geminid meteors observed. Only two violet meteors (both brilliant objects at magnitude -5 and -6) were observed from a total of 9,372 shower meteors. It is possible that the cause of the violet coloration is the ionization of nitrogen in the Earth's atmosphere.

Telescopic Meteors

Some meteor showers are composed of members so dim that they fall well below the limit of naked-eye visibility. Several faint meteor showers – so-called 'telescopic' showers – have been discovered to take place at the end of August and September. One of the most interesting of the year's telescopic showers is active on the night of 11 December and its radiant lies near the star 11 Canis Majoris. From careful observation, it has been found that the members of this elusive shower move in a very elongated orbit around the Sun. At their closest point, the 11 Canis Majorids approach to within 5.6 million km of the Sun – literally just above the Sun's flames and ten times closer to the solar furnace than sweltering Mercury, the innermost planet.

It must be stressed that telescopic meteors are not easy to identify. The rate of faint sporadic meteors is relatively high, and there may well be more faint sporadic meteors than shower meteors. Potentially fruitful watches should ideally be conducted at dark-sky locations well away from city streetlights and on moonless evenings. A rich field telescope with a low-power eyepiece will be necessary to take in as much area of sky as possible. Even reasonably bright meteors within naked-eye range can be difficult to capture on photographic film, and it would be pointless for the novice to attempt to record a faint shower with basic equipment. It is easy to record the image of a dim tenth magnitude star on an undriven time-exposure of, say, five minutes, but a bright second magnitude meteor which appears in the sky for just a fraction of a second may not register at all on the same exposure.

Sophisticated Meteor Detection

Soon after radar was developed during World War II, it was noticed that the trains of ionized gases produced in the Earth's atmosphere in the wakes of meteors acted as strong temporary reflectors for radio waves. The first big radar result was obtained on 10 October 1946, when the Earth passed close to Comet Giacobini-Zinner and ploughed through a trail of debris left in space by the comet; during this period the radar-echo rate of incoming meteors increased by 5,000 times. The first annual meteor shower to be investigated was the Eta Aquarid shower, in May 1947, and the results provided a valuable insight into years of painstaking visual work.

By looking at the way in which radio waves are reflected by meteors, astronomers are now able to determine the position and velocity of meteors accurately. It is possible to obtain such information on meteors which are below the equivalent of visual magnitude 15, a staggering 10,000 times fainter than meteors visible to the naked eye or recorded on conventional photographs. Aside from its sensitivity, the beauty of meteor observation by radar is that it can be performed at any time during the day and in most kinds of weather. Radar has provided an increase in our knowledge of meteors which is comparable in magnitude to the application of triangulation methods to determine meteor heights back in 1798.

3

THE LION ROARS TONIGHT

THE STORY OF THE LEONID METEOR SHOWER

Each November, without fail, the Earth passes through a small portion of a long stream of material deposited in space by Comet Tempel-Tuttle. The multitude of particles which burn up in the Earth's atmosphere appear to radiate from an area of sky in the constellation of Leo, within the sickle shape of stars which makes up the lion's head and flowing mane. The Leonid meteors are visible every year between 15 and 20 November, with most activity occurring on 17 to 18 November.

The Leonids are fast meteors – at a speed of 71km per second relative to the Earth, they are the swiftest meteors of all the annual showers. Many Leonids leave behind them faintly glowing trains that can persist in the atmosphere for a few moments after the meteor's bright flash has disappeared. The normal rate for Leonids lies in the region of 5 to 20 meteors each hour during maximum, occasionally rising to 20 to 50 meteors. But every 33 years or so, the Leonids display much more activity than usual; since they were first recorded nearly 1,100 years ago, there have been some 35 spectacular Leonid displays and 23 Leonid star-storms.

The First Storm

The first recorded Leonid meteor storm burst into life in a most spectacular way. In the small hours of 13 October in AD 902, those who happened to be at large under the dark autumnal night skies of southern Europe and north Africa witnessed a terrifying sight. To their astonished eyes, accustomed though they were to the flash of an occasional brilliant meteor, multitudes of stars seemed to detach themselves from the heavenly vault and fall like rain to the Earth below. Contemporary

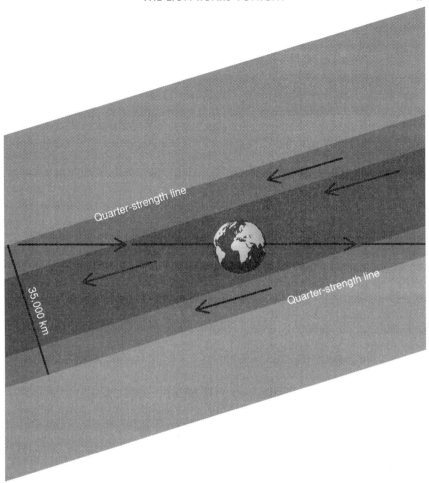

The Earth as it passes through the centre of the Leonid stream.

accounts illustrate vividly the population's fear and consternation that the universe around them had descended into fits of cosmic chaos. A Moorish (Muslim) account quoted by Henry Newton (see Bibliography) notes that 'that night there were seen, as it were lances, an infinite number of stars, which scattered themselves like rain to right and left, and that year was called the Year of the Stars.'

Muslims of the Middle East commonly believed that meteors were benevolent celestial phenomena – balls of fire hurled at terrestrial demons by the angels – and usually regarded them with little fear. But the Leonid storm of AD 902 was an event of such meteoric ferocity that even the most rational observer must have thought that the blazing heavens were a sign of some kind of mighty celestial upheaval directed at humanity. The Christians who saw the event were mindful of the fact that the new millennium was less than 100 years away. What increasingly awe-inspiring displays of heavenly disarray did God have in store for humankind as the second millennium drew nearer? Scriptures mentioned such heavenly phenomena as stars falling to the Earth as signs presaging the end of the world. Christian Europe was deeply immersed in its so-called Dark Ages, a period of scientific and cultural stagnation which made little progress towards enlightenment in the 1,000 years after the decline of the Roman Empire. Confronted with a problem, a western natural philosopher of the Dark Age era would hurry to consult the works of ancient Greek philosophers rather than pursue a new line of reasoning. Dark Age philosophers attempted to define God's reason for the appearance of celestial phenomena and the implications for mankind, rather than to elicit the true nature of heavenly events.

In the fourth century BC, the ancient Greek philosopher Aristotle had gone to extraordinary lengths to catalogue and explain all manner of meteorological and astronomical events. Aristotle's *Meteorologica* (from the Greek *meteoros*, meaning 'lofty'), written around 340 BC, was a very popular philosophical textbook in classical times. At the time of the meteor storm of AD 902, most copies of *Meteorologica* were in Arabic (it was only translated into Latin during the late twelfth century). For those who had access to the book and could read Arabic – and indeed, there were many who could at centres of Muslim learning across North Africa and Moorish Spain – *Meteorologica* was undoubtedly in great demand in the weeks and months following the first Leonid meteor storm, as people clamoured for an explanation for the unprecedented celestial display.

In *Meteorologica*, Aristotle attempts to offer a description and a 'scientific' explanation for everything seen in the skies. The universe was described as being composed of four very basic substances – air, fire, earth and water – and these elements were affected by attributes of dryness, moistness, heat and cold. A flame is hot and dry; water is cold and moist; earth is cold and dry; and air is moist and warm. On the face of it, Aristotle's classification was a neat and sensible scheme which appeared to embody common sense. Each element resided comfortably within its very own sphere of space between the Earth and the Moon, and the perfect harmony of Aristotle's heavens was usually maintained unless the elements were forced to mix with each other; this happened when the Sun's heat intermittently created terrestrial 'exhalations', which got caught up in the busily revolving elemental spheres above. When this took place, the heavens displayed a host of visible phenomena.

Aristotle reasoned that there were two types of exhalation from the Earth – steam, arising from within the Earth and vapour from expanses of water; and dry fire, which was emitted from volcanic eruptions. If an eruption of fire was exhaled by the Earth to the uppermost reaches of the elemental spheres above, then 'immediately beneath the circular celestial motion wherever the conditions are most favourable... this composition bursts into flame when the celestial revolution sets it in motion.' The explosion of fire could cause a 'beam' (presumably a comet), 'torches' (regular meteors) or 'goats' (fireballs) which emitted sparks as they flew across the sky.

The notion that meteors were initiated by the Earth's 'exhalations' and their interactions with the atmosphere took a long time to be replaced by the modern view of an object heating and emitting light as it burned up in the Earth's atmosphere. Sir Isaac Newton speculated in his famous book *Opticks* (1704) that from the Earth 'sulphureous streams, at all times when the Earth is dry, ascending into the air, ferment there with nitrous acids, and sometimes taking fire cause lightning and thunder and fiery meteors.' Newton was hostile to ideas that the solar system contained anything but the Sun and the major planets, their satellites and comets, 'except perhaps some very thin Vapours, Steams, or Effluvia, arising from the Atmospheres of the Earth, Planets, and Comets.' Thus the great natural philosopher ruled out the very existence of meteoroids, preferring the idea of an outer space free to move according to his laws like a clockwork model, unspoiled by any loose flies in the Newtonian ointment.

The Second Storm

The second spectacular Leonid meteor storm on record took place in November in AD 934, 32 years after the first. This was truly a generation later – few people who had seen the star-storm of AD 902 would have been around to witness this event, for these were times when life expectancy was very low, barely exceeding 32 years in tenth-century Europe. Again, the star-storm was witnessed in southern Europe and North Africa, and it was also recorded by thousands of people on the other side of the Earth, in China.

In tenth-century China, the Sung Dynasty ushered in a period of great cultural advancement, where story-telling and poetry became a popular form of entertainment – one can imagine a theatre in Beijing advertising 'An Evening with Shu-Tian-Sta – Poetic Recollections of the Great Celestial Star-Storm'. The ancient Chinese culture had an advanced understanding of the heavens and was far more refined than that of the West; the Chinese Empire was frequently referred to as the Celestial Empire because of its deep reverence for heavenly events. At the time of the Leonid storm of AD 934, the first detailed Chinese maps of the heavens were being prepared. Thousands of instances of sky phenomena are recorded in ancient Chinese annals. Astrologers to the emperor's court were held in the highest esteem, and many resided within the imperial palace itself. Astrologers' advice, based on careful astronomical observation, was greatly respected by the emperor, and helped him to make important decisions about the running of the state. However, should astronomical errors be made, and important events like eclipses fail to be predicted accurately, it was not unknown for the unfortunate bungling stellar soothsayers to meet their doom. One story tells us that two astrologers were executed for failing to predict a lunar eclipse in 2,136 BC accurately.

During the gulf of more than 1,000 years since these times there have been 56 significant displays of the Leonid meteor shower, including no less than 21 events which can justifiably be called meteor storms. Probably the greatest Leonid storm in history took place between 23 and 27 October 1533, peaking at midnight on 25 October over Europe; the magnificent display was also seen in the pre-dawn skies of China, Korea and Japan.

An Eighteenth-Century Star-Storm

The first modern account of a Leonid star-storm was given by Alexander von Humboldt (1769–1859), a famous German naturalist who happened to be on an expedition to South America with his French colleague, the scientist Aimé Bonpland, at the time of the great Leonid storm of November 1799. A fortnight before the meteor storm was due to take place, Humboldt had successfully defended himself against an unfriendly Venezuelan native known as a 'Zambo' who, brandishing a club in one hand and a knife in the other, had attempted to spoil the German's South American vacation. Seemingly unperturbed by this attack, the scientists' observations of the next day's solar eclipse proceeded as they had planned.

The mighty Leonid meteor storm took place on the morning of 12 November. From the balcony of their quarters in the Venezuelan village of Cumana, some 300km east of Caracas, the two scientists made their observations. In his account (see Bibliography), Humboldt wrote:

> At about 2.30 the most extraordinary, luminous meteors began rising out of the sky from the east and northeast... almost all the inhabitants of Cumana witnessed these phenomena because they had left their houses before four o'clock to attend early morning mass. They did not behold these fireballs with indifference; the oldest among them remembered that the great earthquakes of 1766 [33 years earlier] were preceded by similar phenomena.

Humboldt remarked that within a quarter of an hour of the storm beginning the sky was full of meteors – crowded so much that he reported 'thousands of meteors and fireballs moving regularly from north to south with no parts of the sky so large as twice the Moon's diameter not filled each instant with meteors.' Many were as bright as Jupiter and emitted showers of sparks as they flew across the sky, and some exploded violently in a spectacular terminal flare. Incredibly, Humboldt reported that the storm was visible in the morning skies half an hour after local sunrise.

Deeply fascinated by what they had witnessed, Humboldt and Bonpland set about collecting as many reports of the 1799 storm as they could find. They discovered that the Leonid star-storm had been widely observed in South America; they also found an account by Moravian missionaries (quoted by Henry Newton – see Bibliography) as far north as Labrador and Greenland, which stated:

> On the 12th of Nov, there was at Nain and Hoffenthal a strange appear-
> ance in the air, which greatly frightened the Eskimos. For there fell
> down to earth in the four quarters of the heavens, about daybreak, very
> many fireballs, some of which seemed to be half an ell [about 0.5m] in
> diameter. This phenomenon was at the same time seen at New Herrnhut
> and Lichtenau in Greenland.

Andrew Ellicott, an American government official, observed the 1799 storm from the deck of a ship off Cape Florida in the Gulf of Mexico. He recorded the event in his journal (see Bibliography):

> November 12th, 1799, about three o'clock AM. I was called up to see the
> shooting of the stars (as it was commonly called). The phenomenon was
> grand and awful; the whole heavens appeared as if illuminated with sky-
> rockets, which disappeared only by the light of the Sun after daybreak.
> The meteors, which at any one instant of time appeared as numerous
> as the stars, flew in all possible directions, except from the earth, toward
> which they all inclined more or less; and some of them descended
> perpendicularly over the vessel we were in, so that I was in constant
> expectation of their falling upon us.

Nineteenth-Century Storms

Thirty-three years later, as accounts of the 1799 meteor display were beginning to sound like fanciful exaggerations, the skies hosted another great Leonid storm. It was during this event, on the night of 12 November 1833, that the concept of a radiant – a single small area of the sky from which meteors appear to emanate – became apparent. The 1833 storm sparked modern interest in the scientific study of meteors. American observers described the meteors as being as thick as snow coming down in a snowstorm. At Greenwich in London, eight observers were each allocated an area of sky, allowing complete coverage; a total of 8,000 meteors was recorded, with a maximum of 4,860 falling between 1am and 2am.

The Victorian science writer and astronomer Agnes Mary Clerke (1842–1907) researched accounts of the event thoroughly in order to describe the storm which had taken place nine years before she was born. In her *History of Astronomy in the Nineteenth Century* (1885) she writes the following:

On the night of November 12–13, 1833, a tempest of falling stars broke over the earth. North America bore the brunt of the display. From the Gulf of Mexico to Halifax, until daylight put an end to the display, the sky was scored in every direction with shining tracks and illuminated with majestic fireballs. At Boston, the frequency of meteors was esti-mated to be about half that of flakes of snow in an average snow-storm. Their numbers, while the first fury of their coming lasted, were quite beyond counting; but as it waned, a reckoning was attempted, from which it was computed, on the basis of that much-diminished rate, that 240,000 must have been visible during the nine hours they continued to fall.

In his account (see Bibliography), Professor Denison Olmsted of Yale University claimed that the early morning of 13 November 1833 was:

...rendered memorable by an exhibition of the phenomenon called shooting stars, which was probably more extensive and magnificent than any similar one hitherto recorded.... The firmament was unclouded; the air was still and mild; the stars seemed to shine with more than their wonted brilliancy.... Probably no celestial phenomenon has ever occurred in this country... which was viewed with so much admiration and delight by one class of spectators, or with so much astonishment and fear by another class. For some time after the occurrence, the 'Meteoric Phenomenon' was the principal topic of conversation in every circle.

The Leonid meteor radiant.

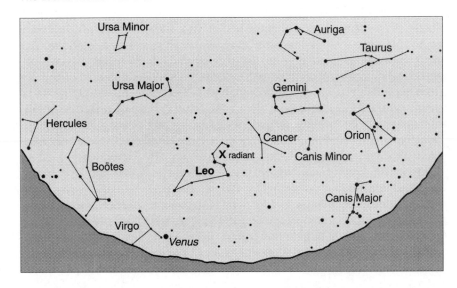

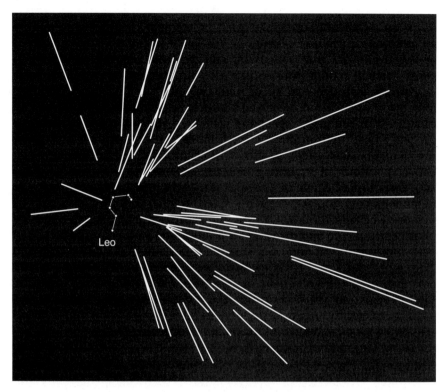

Leonid storm of 1866. The plots of the bright meteors are based on an observation by Professor Harkness of the US Naval Observatory.

Olmsted and others observed that the Leonid radiant remained fixed within the 'sickle' of the constellation of Leo over a period of several hours during the storm, in spite of the fact that the Earth had rotated to the east through at least 30 degrees. This led Olmsted to the correct conclusion that the meteors must have been coming from an inter-planetary rather than an atmospheric source – a conclusion which, sur-prisingly, was not endorsed by some of the leading astronomers of the day. Olmsted reported that one observer had actually detected the Leonid stream before the meteor storm had commenced, purely by the reflection of sunlight from the dense meteoric clouds at the stream's heart.

A South Carolina planter (quoted by G. F. Chambers – see Biblio-graphy) vividly described the 1833 star-storm:

I was suddenly awakened by the most distressing cries that ever fell on my ears. Shrieks of horror and cries for mercy I could hear from most of the Negroes of the three plantations, amounting in all to about 600 or 800. While earnestly listening for the cause I heard a faint voice near the

door calling my name. I rose, and taking my sword, stood at the door.
At this moment I heard the same voice still beseeching me to rise, and
saying, 'Oh my God, the world is on fire!' I then opened the door, and
it is difficult to say which excited me most – the awfulness of the scene,
or the distressed cries of the Negroes. Upwards of 100 lay prostrate on
the ground, some speechless, and some with the bitterest cries, but
with their hands raised, imploring God to save the world and them.
The scene was truly awful, for never did rain fall much thicker than
the meteors fell towards the earth; east, west, north and south, it was
the same.

As predicted, 1866 saw another magnificent Leonid star-storm. Sir Robert Ball, an eminent Victorian astronomer and popularizer of astronomy,
gives an account of the event in his classic *Story of the Heavens* (1890):

Such was the occurrence which astonished the world on the night
between November 13th and 14th, 1866. We then plunged into the
middle of the [Leonid] shoal. The night was fine; the Moon was absent.
The meteors were distinguished not only by their enormous multitude,
but by their intrinsic magnificence. I shall never forget that night. On
the memorable evening I was engaged in my usual duty at the time of
observing nebulae with Lord Rosse's great reflecting telescope. I was of
course aware that a shower of meteors had been predicted, but nothing
that I had heard prepared me for the splendid spectacle so soon to be
unfolded. It was about ten o'clock at night when an exclamation from
an attendant by my side made me look up from the telescope, just in
time to see a fine meteor dash across the sky. It was presently followed
by another, and then by others in twos and in threes, which showed
that the prediction of a great shower was likely to be verified. At this
time the Earl of Rosse joined me at the telescope.

The two astronomers made their way to the top of one of the high walls
that surrounded the telescope and chose to view the event from there:

...for the next two or three hours we witnessed a spectacle which can
never fade from my memory. The shooting stars gradually increased in
number until sometimes several were seen at once. Sometimes they
swept over our heads, sometimes to the right, sometimes to the left,
but they all diverged from the east... all the tracks of the meteors
radiated from Leo. Sometimes a meteor appeared to come almost
directly towards us, and then its path was so foreshortened that it had
hardly any appreciable length, and looked like an ordinary fixed star
swelling into brilliancy and then as rapidly vanishing. Occasionally
luminous trains would linger on for many minutes after the meteor
had flashed across, but the great majority of the trains in this shower

were evanescent. It would be impossible to say how many thousands of meteors were seen, each one of which was bright enough to have elicited a note of admiration on any ordinary night.

The astronomical world, having been convinced by the storms of 1799, 1833 and 1866 that the Leonids had a 33-year cycle, awaited the night of 14 November 1899 with great expectations. The interest of the general public, too, had been greatly aroused by newspapers, keen to promote the forthcoming Leonid shower as an unmissable display of celestial pyrotechnics. However, calculations made earlier by the British astronomers G. Johnstone Stoney and A. M. W. Downing suggested that the densest part of the Leonid swarm would not be encountered by the Earth on this occasion. It was shown that the meteoroid swarm had been gravitationally deflected by Saturn in 1870 and perturbed by a close approach to Jupiter in 1898, and unless the swarm was very wide, a major meteor display was unlikely to occur on this occasion.

But few people took any notice of these negative predictions, and there was great disappointment when the anticipated Leonid star-storm of 1899 failed to happen. In his account (see Bibliography) Ellison Hawks, a British astronomy popularizer, summed up the situation well:

Sometimes when great displays of meteors are expected, they do not take place, as for instance on 13th November 1899. A great shower of shooting stars was expected on that date, and a crowd of people assembled not far away from where I was observing. They waited expectantly for many hours, but the shooting stars did not come. Still the people waited, till it was long past midnight, and then some University students, who intended to have some fun, set off a great flight of sky-rockets – the sort which burst and send out a shower of coloured stars. When the people saw all these beautiful 'firework-stars' they cried: 'Look! look! There are the shooting stars!' After the fireworks had died away the people went home, quite thinking that they had seen the real shooting stars, and I do not believe that they know to this day that what they saw on that night were only fireworks, set off by the University students!

The 1899 Leonids did put on a display, albeit a much-muted one, but they failed to impress most of the public who made the effort to observe them, which isn't surprising in the light of the thorough pre-event hype in the pages of the world's press. The disappointment was described by American meteor expert Charles Olivier as 'the worst blow ever suffered by astronomy in the eyes of the public, and has indirectly done immense harm to the spread of science among our citizens' (see Bibliography).

Comparable levels of pre-event publicity were allocated to the in-famous Comet Kohoutek, discovered by the Czech astronomer Lubos

Kohoutek at the Hamburg Observatory in March 1973 and predicted to become so bright as to be 'the comet of the century' (see page 30). In fact, the comet barely reached naked-eye visibility, although it was a thoroughly worthwhile object in the eyes of most amateur and professional astronomers; the third US Skylab crew in Earth orbit spent 10 per cent of their experiment time on studies of Kohoutek. Not surprisingly, the public were disappointed that it failed to live up to expectations; this 'damp squib' left a bitter taste in the mouths of those led to expect a lot more than just a dull, nebulous patch of light. Similarly, media coverage of the 1985–6 apparition of Halley's Comet tended to give the impression that a grand celestial spectacle would be enjoyed by all (see page 24). To digress slightly, it can be revealed here for the first time that in December 1985 a Central TV film crew in England faked the image of Halley's Comet for their news report because the comet, as seen through the eyepiece of a 175mm refractor at an observatory in Edgbaston (near the centre of Birmingham), was far too dim to register. I should know, because I colluded in this benign deception, even going so far as to suggest that they should add a little distortion to the faked image to simulate atmospheric shimmer! The televised image of Halley's Comet which millions briefly saw that evening was nothing more than an underexposed lightbulb in the observatory! Like those Victorians who saw the fireworks masquerading as Leonid meteors, the curiosity of late twentieth-century comet viewers was at least partly satisfied.

Sir Robert Ball, expert witness of the 1866 Leonid storm, gave a series of lectures around Britain to promote the 1899 Leonids. The *West Bromwich Free Press* (24 November 1899) gives an amusing account of a talk Ball gave a week after the big disappointment:

> Among those present were the Rev G. M. Bacon and his daughter, who made a perilous [nocturnal!] balloon voyage in quest of the shooting stars. Miss Bacon, whose arm was fractured in the descent near Neath, received an ovation as she entered the hall.... Sir R. Ball said that during the past week a considerable number of the community had lost more or less time looking out for the shooting stars, which it must be honestly confessed had not come (laughter).
>
> Astronomers expected them to a certain extent, but no one predicted their arrival with any certainty. They might have been seen in other parts of the globe. The shower of shooting stars would only pass for a few hours across the critical point, so to speak, and it was but the merest chance that we should be on the fortunate side of the Earth to witness the display. Unfortunately we were not (laughter).
>
> The chance of seeing the meteor shower this year was over, but there would be another chance next year. If they were disappointed again next year, they would have to wait for 30 years before they would get another chance (laughter).

Ball was right in that the best of that year's Leonid shower may have been observed in other parts of the globe, especially in far eastern Asia; Chinese observers reported a good burst of activity on the morning of 15 November 1899. The Leonids of 1900 and 1901 proved to be more spectacular than anyone had hoped for, given the disappointment of the previous year, and they were well observed in parts of the US and Canada. However, public interest in scientific forecasts of a bumper meteor harvest was nowhere near as high as it had been in 1899. But in 1900, rates exceeding 1,000 meteors per hour are alleged to have caused much consternation in isolated rural communities near Hudson Bay, Canada, leading to something of a panic. The Leonids of 1901 briefly attained rates as high as 2,000 per hour, defying watchers' attempts to record each meteor individually.

The 1966 Star-Storm

In November 1966, it was announced that the world's fishing fleets had made their biggest annual catch in history. That month, the Earth itself would trawl the oceans of space to make a record catch of meteors from the dense Leonid shoal. On 1617 November 1966, the Leonids put on the twentieth century's greatest display of shooting stars, as meteors fell 'as thick as snowflakes'. The storm was visible in the morning skies between western parts of the US and eastern Asia. The storm reached its peak at 12:00 UT on 17 November, when the rate was estimated to be about 150,000 naked-eye meteors per hour. It was calculated that the meteors passed over Arizona at a rate of 2,300 each minute in a period of 20 minutes from 5am local time on 17 November. Observers in 1966 commonly reported the illusion of travelling rapidly through a tunnel lined with stars. No doubt a future Leonid storm will give those fortunate viewers familiar with the television series *Star Trek – The Next Generation* the impression of peering at the approaching galactic star-fields from the USS Enterprise's 10-forward lounge at high warp speeds!

The manned spacecraft Gemini 12 had been launched into Earth orbit a week before the 1966 Leonid storm. On 12 November, astronauts Edwin 'Buzz' Aldrin and James Lovell, flying 290km above South America, observed a total eclipse of the Sun. One hundred and sixty-seven years before, two very different explorers had observed the solar eclipse of 28 October 1799 from South America – the intrepid von Humboldt and Bonpland, who also saw the amazing Leonid storm a few days later. But for safety reasons, mission control had decided to bring Gemini 12 back to Earth before the densest part of the Leonid stream was encountered. The astronauts had been fully briefed before the flight and were well aware of the hazards which cosmic debris presented to them. On the mission's third day, Aldrin performed a space-walk out of the cramped Gemini cabin and made his way across to the Agena craft with which they had docked on their third Earth orbit. Reaching over the equipment bay of the Agena, Aldrin exposed a micrometeoroid collection panel on the craft, an experiment to sample the interplanetary dust in near-Earth orbit and gather important data on it. Had Gemini 12 stayed in orbit just a day and a half longer to witness the arrival of the Leonid storm, and had the craft been in a position to observe the event above the Earth's dark side, then a fabulous, scintillating display of countless thousands of meteor flashes would have been seen below them.

OBSERVING THE LEONID STAR-STORM

By explaining how to observe the long-awaited Leonid star-storm, due to take place in one or more of the closing years of the twentieth century, I am faced with a difficult task – although I am familiar with meteor showers, I have never seen anything like a meteor storm. I have seen one extremely rare event, namely the grazing occultation of the star 44 Capricorni ZC3177 by the Moon's limb. On that occasion, I was completely unprepared for the phenomenon that I actually observed – an unexpectedly complicated display of fadings and brightenings which took place in such rapid succession that there was no way I could capture it adequately by means of a stopwatch and notebook. A tape recorder would have been invaluable; better still, a real-time video image of the grazing phenomenon may have been sufficient to determine the degree of fadings and brightenings, so that a configuration of the presumed multiple star system that caused the brightness fluctuations could be inferred.

Each observer will be expecting to do their own thing on the night – whether it is just to look in wonder at the star-storm (as I'm sure many professional astronomers will), to try to record part of the event as a souvenir, or to conduct a proper scientific study. It's a little like any spectacular naked-eye astronomical event which takes place in a compressed period of time, say a lunar eclipse – an event which anyone can enjoy, whether or not they are deeply involved in astronomy on an amateur or professional basis. Faced with a massive bombardment of space dust, it is certain that some observers will drop all their carefully laid plans and simply gawp in utter astonishment for hours on end. Some – although hopefully nobody who has read this book – may even be frightened by the Leonid storm, just as some unenlightened people around the world are terrorized by a total eclipse of the Sun.

Leonid Storm Forecasts to 2002

The observability of a frenetic burst of meteor activity that lasts for one or two hours, with rates exceeding 1,000 meteors per hour, depends on whether the meteor radiant is visible in the sky at the crucial time of maximum activity, although it is possible that a spectacular Leonid starstorm can produce a good display even when the radiant is below the eastern horizon and has yet to rise.

In mid-November each year, from Britain, the Leonid radiant, located in the sickle-shaped asterism of Leo, the lion's head, rises in the eastern sky at around 23:00 UT and culminates due south at around 6:30, an hour before sunrise. An ideal set of circumstances guaranteed to produce a spectacular storm would be for the Earth to cross the plane of Comet Tempel-Tuttle's orbit (along which lies the Leonid meteoroid stream) within weeks of a close approach to the comet itself. If this took place in a clear moonless sky around 04:00 on 17 November, British observers would be in for an unrivalled display of celestial pyrotechnics. In reality, there are mixed prospects in the predictions that have been made for the Leonid maxima occurring from now until the year 2002.

From Europe, the maximum of 1997's Leonids was not observed visually since it took place during the afternoon of 17 November (literally as this book was being prepared for press). Visual observers had to contend with the glare of a waning gibbous Moon that threatened to drown out all but the brightest meteors seen in the pre-dawn skies. Nonetheless, enhanced rates were clearly detected by visual meteor observers in the Netherlands, who estimated a rate of 30 to 40 meteors per hour, including one brilliant blue fireball of magnitude -6.

But 1997's real Leonid action was seen from the western US, where

The Earth's passage through the Leonid meteoroid stream near Comet Tempel-Tuttle, showing the intensity of observed meteor showers through history.

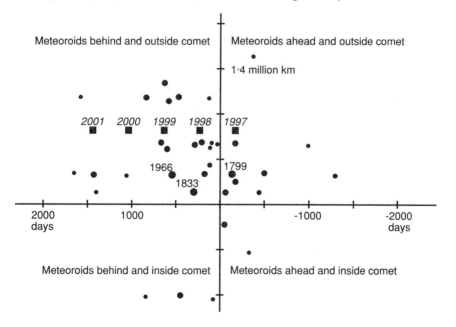

between 12:00 and 12:30 UT (4am to 4.30am local time) dozens of brilliant fireballs were observed. Reports from southern California told of a splendid burst of activity, in which one fireball of magnitude -4 or brighter was seen almost every minute, some of these equalling the brilliance of the Moon riding directly above the observers! Many meteors displayed beautiful luminous trains that lasted several minutes. This superb level of Leonid activity strongly indicates that the displays of 1998, 1999 and 2000 will live up to our high expectations.

On 17 November 1998, the Earth's passage through the dense parts of the Leonid shoal occurs at 19:45, and the meteor activity on this occasion will be most favourably observed from Asia. The Moon will be nearly new and well out of the way, and there is a strong possibility of an outstanding shower and possibly a storm, with predicted rates of up to 5,000 meteors per hour. From Britain, it might be possible to observe enhanced meteor rates – perhaps higher than 100 per hour – as the radiant nudges above the eastern horizon.

Of all the predictions, the Leonid display forecast for the early morning of 18 November 1999 has the greatest chance of being a true meteor storm, comparable with those incredible events already described in this chapter. Moreover, the maximum takes place at 01:50, meaning that from Britain the radiant is located more than 20 degrees above the eastern horizon during the peak time of the storm – higher, if one travels further east into central Europe, Africa or Asia. The Moon is only nine days old in Aquarius on this date, and sets at 00:45, long before the starstorm is due to burst forth against the canvas of space.

The following year's Leonid display, with its maximum predicted to occur at 08:05 on 17 November 2000, will be at its best when observed from eastern parts of North America, South America and the South Atlantic. Observers in Britain may detect signs of an increase in Leonid activity in the hours before sunrise, but the event is spoiled somewhat by the presence of a 20-day-old waning gibbous Moon in the constellation of Cancer, 20 degrees to the right of the radiant. Predictions are that there will be an outstanding shower on this morning, possibly a storm; although ideal rates are expected to go beyond 1,000 per hour, the numbers actually observed will be less owing to the bright Moon.

Leonid activity in the years 2001 and 2002 is likely to be much lower than in the preceding years, with little chance of a star-storm. Neither event will be well placed from Britain; in 2001, the maximum takes place at 14:24 over North Pacific skies, and in 2002, the radiant will be situated below Britain's eastern horizon at Leonid maximum at 21:36.

How to Prepare

November nights in the northern hemisphere are usually rather chilly, and from Britain sub-zero temperatures are to be expected at this time of the year. Wrap up well, whether you are planning to observe for an hour or to remain under the stars for the entire night. Thermal longjohns (or tights worn under trousers) are advisable, as are a thick pair of socks, woollen gloves (fingerless, if you plan to write or operate equipment) and a woolly hat to prevent heat escaping from your head. A pair of jeans or thick trousers and a woollen jumper overlaid with a good winter coat is essential, and a scarf will help keep those freezing gusts of air from chilling your neck.

A flask of warm coffee or soup will help maintain your core body temperature, and some confectionery will keep your energy levels up. It is not a good idea to take alcoholic beverages either before or during your observing session, since alcohol only makes the body feel warm for a short period by dilating the blood vessels and increasing circulation in the outer layers of the body. In fact, this process makes a person lose heat rapidly and feel drowsy; to fall asleep in the open in near-freezing temperatures is a dangerous invitation for a hypothermic coma to set in. If you drink, save the alcohol for the celebrations after the event.

If a meteor storm does happen, then an ordinary camera loaded with regular 200 ASA film, set up pointing towards the direction of the radiant and left with its shutter open for a good length of time (say ten minutes) will record all meteors brighter than the first magnitude (see page 121). Today's hand-held camcorders are good enough to record bright meteors, and if you are planning to attend a star party during the Leonid storm, a video record of the event would make a great souvenir.

4

METEORITES: SOUVENIRS FROM SPACE

BOMBARDMENT FROM ABOVE

The l'Aigle meteorite fall in April 1803 finally convinced scientists that stones do occasionally fall to Earth from the depths of space. Ever since, astronomers have realized that the Earth lies prone to cosmic assault. For all our long-range detection capabilities and technological prowess, it must be admitted that, even at the dawn of the twenty-first century, we are still utterly defenceless against bombardment from outer space.

Small meteoroids, such as the majority of objects that comprise the known meteor streams, vaporize entirely in the atmosphere and present no great threat to the Earth or the well-being of its inhabitants. Nevertheless, a full-blown meteor storm in which the heavens are set alight for a few hours by countless bright meteors is entirely capable of inspiring no mean degree of fear in even the most rational of observers!

Meteoroids larger than the size of a grape usually burn up in the Earth's atmosphere and appear as spectacular fireballs. They may be composed of resilient material and have enough mass to withstand their friction-superheated passage through the atmosphere, surviving all the way down to the ground to put a small dent in something. Once a meteoroid has ended its journey through the atmosphere it is classified as a meteorite, and because three-quarters of our planet's surface is covered with water, the majority of meteorites plop unceremoniously into the sea, never to be seen again.

Large meteoroids often explode violently in flight and the fragments distribute themselves over a sizeable portion of terrain. A typical daytime meteorite fall took place on the morning of 10 February 1896, when the skies above Madrid in Spain hosted the sudden appearance of a brilliant fireball. At 9.30am local time, in bright sunshine, the flash of the fireball's explosion was so great that it illuminated the interior of houses. An amazing clap of thunder was heard afterwards – the sonic boom created by the object's supersonic velocity – and concerned members of the local population believed that an explosion of dynamite had occurred nearby. The fireball burst at a height of 24km and was seen as far as 740km from Madrid. As if to commemorate the event in style, pieces of the object landed immediately in front of the National Museum and presented themselves as instant science exhibits, fresh from interplanetary space and too hot to touch after their fiery passage through the atmosphere.

Meteoroids with a mass of less than 10 tonnes are slowed by friction through the atmosphere sufficient for them to lose all their original interplanetary velocity and land on the Earth under the influence of gravity, as if they had simply been dropped from the height of a low-flying aircraft. On New Year's Day 1869, several stone meteorites weighing up to 1kg were recovered on the surface of a frozen lake at Hessel, near Uppsala in Sweden. Their location on a relatively thin layer of ice proved that the objects had not landed with enormous velocity.

Small meteoroids can hit the ground with the same magnitude of force as if they had been thrown by a fast bowler – not nearly as fast as a bullet, but potentially lethal, nonetheless. Any meteoroid with a mass of more than 10 tonnes will hold on to some of its original breakneck interplanetary velocity, penetrating the atmosphere with ease and crashing to the ground with a force capable of excavating a sizeable crater.

TYPES OF METEORITE

Meteoritic material from outer space can be classified as belonging to three main groups, a classification of types first proposed in 1863 by Nevil Maskelyne, Britain's fifth Astronomer Royal. They are: iron meteorites, stone meteorites and a hybrid group of stony-irons. Respectively, Maskelyne referred to these groups as Siderites, Aeroliths (Aerolites) and Mesosiderites (Siderolites); these are designations often used today, and each of these groups has its own divisions and subdivisions right down to individual meteorites themselves.

The total number of stony meteorites in museum collections is far greater than the number of irons and stony-irons combined, the stones outnumbering the others by something like ten to one. Most of the meteorites people have purposely hunted down after they have been seen to fall come under the stony category. But the majority of meteorites just stumbled across in the field by chance, whose fall was not seen (perhaps having lain on the ground for a number of years before their discovery) belong to the iron group.

This blatant discrepancy can be attributed to a simple matter of identification. Stony meteorites tend to blend into most environments well, because they weather quickly and camouflage themselves as just another mundane-looking grey rock. Irons, on the other hand, erode much more slowly than stones and appear noticeably different to the surrounding rock; they are therefore much more likely to be regarded as material alien to that environment, an identification which is obvious even to the eyes of non-geologists.

Over two centuries of the scientific study of meteorites has left behind a convoluted trail of nomenclature, which has developed as scientific knowledge of the composition of meteorites has improved through advances in chemistry, mineralogy and microscopy. Unless they have been derived from the same meteoritic fragment as it exploded in mid-air, no two meteorites have ever been found to be exactly alike in their composition, and this fact is particularly relevant to non-iron meteorites. Adequate classification is therefore made even more difficult, and meteorite experts sometimes find it far more convenient to refer to an individual meteorite by its name (usually given after the nearest town or village to the fall site), rather than by the meteorite's type.

Irons

Iron meteorites can be classified according to the proportion of nickel they contain. The average composition of an iron meteorite is 92 per cent iron, 7 per cent nickel and the remainder of other substances, including cobalt. Specimens of irons are often cut into sections, etched with acid and polished; the result is a shining face crossed with a series of lines at angles to each other, much like a masterpiece of cubist artwork. This unusual appearance is called the Widmanstatten structure, after the scientist Aloys von Widmanstatten who observed and described the phenomenon in 1808, and brings out bands of a nickel-rich mineral called Kamacite. Iron meteorites are graded according to the width of these bands, from fine to coarse, which varies according to the nickel content, but the test only works on meteorites containing less than 12 per cent nickel.

Stones

Stony meteorites are composed of silicates – compounds of metals, silicon and oxygen which make up half of all the known minerals – and are of two main types, namely chondrites and achondrites. Chondrites are so-named because they contain small chondrules (crystallized spheres) of minerals such as pyroxene, while the achondrites contain no chondrules and are an intermixture of various minerals, mainly olivine. The most interesting group of chondrites is the carbonaceous variety, in which many of the basic building blocks of life have been found, including hydrocarbons, fatty acids and amino acids.

Stony-Irons

Stony-irons are divided into three groups, according to the types of minerals which are associated with the nickel-iron alloy found within the meteorite in question. The most abundant type of stony-iron is pallasite, a nickel-iron alloy mixed with equal proportions of olivine, of which 39 examples are known. Mesosiderites, with 32 known specimens, are the next most populous group of stony-irons, and are composed of a mixture of nickel-iron and silicate material. Lodranites, the rarest of the stony-iron meteorites, have bronzite mixed with nickel-iron or olivine, and only two examples are known.

Microscopic diamonds have been discovered in a number of meteorites, including the Allende carbonaceous chondrite (Mexico, 1969), the Murray and Murchison carbonaceous chondrites, and the Indarch (Azerbaijan) chondrite. These meteorites have been found to contain higher than usual amounts of the noble gases xenon and krypton, suggesting that they may have come from outside the solar system where there was a different mix of surrounding gases at their birth. It is thought that diamonds can be formed at low pressure; perhaps these tiny specimens condensed around microscopic carbon grains which were bathed in the hot gases streaming from an ancient red giant star.

THE ORIGINS OF METEORITES

Detailed studies of meteorites have enabled scientists to theorize about how they are likely to have originated. It is generally accepted that most meteorites are small fragments of large ancient asteroids which have been mechanically broken up by cosmic collision.

When the solar system came into being some 4.6 billion years ago, pulling itself together out of a gravitationally collapsing nebula of gas and dust, countless small solid bodies soon coalesced. The big objects in the young solar nebula – the Sun and its attendant planets and their satellites – rapidly swept up most of these solid bodies, but a few avoided being captured. Some of these asteroidal bodies, even though they may have been no larger than 500km in diameter, were large enough to develop their own internal heating as a result of radioactive decay, perhaps combined with some heating as a product of electric currents induced by the hot plasma streaming from the active young Sun.

Silicate material in these asteroidal bodies underwent partial melting and formed the material of most of the meteorites – the stones, irons and stony-irons. Rudimentary iron-rich cores, overlain by a differentiated structure of progressively lighter materials, settled out in their hot molten mantles, later cooling and crystallizing. The material of the pallasite stony-irons is believed to have formed at the boundary of the core and silicate layer of a large parent body. Many of these large asteroids experienced high-speed collisions with meteoroids and asteroids, further heating and shock-metamorphosing some of their material, and the fragments were distributed throughout space as meteoroids.

Most known asteroids, including 930km-diameter Ceres (the largest asteroid), are very dark objects which may be of the same general composition as the carbonaceous chondrite meteorites. These 'C' type asteroids, comprising three out of four asteroids in the main belt, are very ancient bodies whose composition is considered to be similar to the original solar nebula out of which the planets developed 4.6 billion years ago. Although they themselves have been bombarded by meteoroids and other asteroids during their long history, they are preserved more or less intact and have not undergone extensive internal melting or reheating after their formation. If sampled directly by spacecraft, such bodies might represent veritable Rosetta Stones of the solar system, enabling scientists in the future to gain a deeper understanding of what our cosmic neighbourhood was like shortly after its formation.

'S'-type asteroids make up around 15 per cent of known asteroids, and their composition matches the stony-iron meteorites. 'M'-type asteroids are composed of an iron-nickel alloy and are likely to be the cold naked cores, or fragments of cores, of once larger asteroids, whose mantles have been chipped off by impact processes. These vast ingots of pure metal may be exploited in the future by being towed into Earth or lunar orbit from the asteroid belt and mined for their precious content.

It would be an understatement to say that trying to piece together the history of meteorites is a complicated process, but the clues are now beginning to be understood in their correct context. A significant leap in our understanding of these visitors from space – equivalent to that experienced by lunar science following the Moon landings – will occur when asteroids are studied by dedicated spaceprobes in the twenty-first century, and pieces of them are returned to Earth for analysis.

ASSAULT FROM THE SKIES

British Meteorite Falls

There have been 24 recorded meteorite falls in the British Isles since the seventeenth century – eight in Ireland, three in Scotland, two in Wales and eleven in England. This distribution is a fair meteoritic sprinkling, considering the surface area and population of each country.

The last recorded fall took place on 5 May 1991 in the village of Glatton, near Peterborough in Cambridgeshire. As he was spending a quiet Sunday tending his onions, retired civil servant Arthur Pettifor was startled to hear a 'very loud whistling, whining, screaming noise'. Shortly afterwards, the meteorite – a stony chondrite containing 23 per cent iron and 5 per cent nickel-iron, 7cm in diameter and weighing 600g – fell from an overcast sky and landed in a nearby hedge. The Glatton event was the first British meteorite fall for over a quarter of a century.

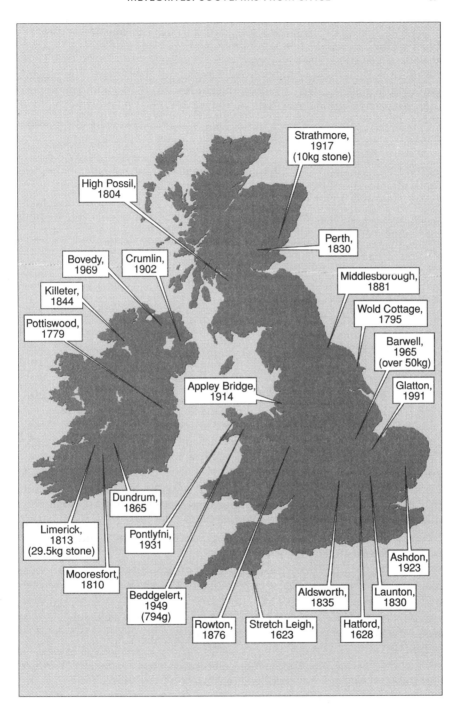

Significant UK meteorite falls.

The most significant British meteorite fall of modern times took place in the sleepy village of Barwell in Leicestershire on Christmas Eve 1965. Festive preparations were rudely interrupted at about 4.20pm local time, when more than 50kg of rock rained down upon the village. The object's passage through the early evening skies was seen by many in the region, despite variable cloud cover. The meteorite passed over eastern Coventry, some 23km to the southwest of Barwell, and was reported to be orange-red in colour, giving off sparks and sporting a tail 20 degrees long.

It was estimated that around half of Coventry's population of 300,000 heard the sonic effects created by the meteorite's passage, which was reputedly loud enough to have shaken buildings and shattered light-bulbs. One man likened the event to a dive-bomber attack he had experienced during World War II! The frequency range of the sounds emitted by the object may have been too high for the human ear to register entirely, as evidenced by the bolting of horses shortly before the meteorite made its visual and auditory presence known to humans. Police were inundated with callers concerned that they had seen a flying saucer or heard the explosion of a bomb

The skies were completely overcast in Barwell that afternoon. One resident, Ernest Crow, was returning home from work when he noticed a sudden flash in the sky accompanied by a loud bang. Seconds later, Crow took cover against a nearby wall as an object 'swished' down from the sky and landed near him with a thud. Almost immediately afterwards, a number of similar objects impacted nearby in rapid succession. One hit the road, shattering and breaking his neighbour's window.

The hand-sized crater the object had made in the road was filled with granules and surrounded by a fine white dust. Crow's neighbour, Joseph Grewcock, emerged from the house to see what had broken his window, and found several small rock fragments which were too hot to handle. Another fragment of meteorite had entered his house and come to rest in a vase of flowers, where it was discovered a week later. Some distance away at a knitwear factory (whose employees had been on holiday at the time), a 1kg stone had smashed through an asbestos roof and ended up beneath floorboards in a work area.

Once the national news media got hold of the story and had run with it several days later, souvenir hunters flocked to Barwell and its environs to search for meteorite fragments. There were differing opinions as to the nature of the event. One local man was quoted as saying: '...if it was a meteorite, it fell out of season – they usually fall in November or January'(!).

A collection of meteorite fragments was amassed by local Police Constable George Scott, and geologists soon confirmed that a stony meteorite had indeed landed. In February, the British Museum offered a reward of ten shillings per ounce (28g) for stones weighing more than two pounds (1kg). Over the next four months, some 46kg of meteorite fragments was collected – including the largest fragment found, a piece weighing 8kg which earned the finder the grand sum of £140.

The Barwell meteorite was found to be the most common type of meteorite, a stony chondrite. The individual fragments did not have substantial crusts of fused material caused by frictional heating, and therefore the original meteorite must have broken up at a low altitude. The mass of the original Barwell body was estimated to have been about 90kg – 150 times bigger than the Glatton meteorite of 1991 and about the size of an average VDU. It fell just a day or so after the peak of the Ursid meteor shower but, like the fall of the Mazapil iron in Mexico during the Andromedid shower of November 1885, and the Glatton meteorite just after the maximum of the Eta Aquarid shower, this may be just a coincidence.

Danger!

On the basis of calculations made by Ian Halliday (et al.) of the Herzberg Institute of Astrophysics in Ottawa, Canada, some 16 buildings around the world are likely to be damaged by meteoritic impact each year. The research also predicts that at least one person should be hit by a meteorite every nine years, with the occasional fatality. Meteorite expert Halliday estimates that each year around 5,800 meteorites weighing at least 100g hit the ground. Since the Earth's surface is 75 per cent water, presumably more than 15,000 meteorites splash into rivers, lakes and oceans and are never seen again. These calculations have been based on nearly a decade of research using a network of 60 sky-pointing cameras spread over western Canada. However, the actual number of meteorogenic human injuries across the globe is much lower than Halliday predicts – perhaps a proportion of people who are struck by meteorites attribute their injuries to stones thrown by human hands, and the true cause of their injuries goes unsuspected.

When the Book of Joshua (10:11) tells us 'The Lord cast down great stones from heaven upon them... and they died...', this may allude to an actual fall of meteorites which caused terrible destruction in the Holy Land before the seventh century BC. Two people certain to have been familiar with this biblical passage were reputedly killed by falling meteorites. On 14 September 1511, a Franciscan monk was killed by a small shower of meteoroids in Cremona, Italy, and another monk died after being hit by a meteorite in Milan in 1650. A meteorite that fell on 16 January 1825 in British India is said to have killed a man and seriously injured a woman.

In the mid-seventeenth century, two sailors were allegedly killed when a 4kg meteorite fell on the bridge of their ship. Somewhat luckier were the crew of the barque *Gemsbrok*, whose captain reported a near-fatal encounter with a possible meteoroid. On 9 October 1882, during a southwesterly gale and a thick snow squall, a ball of fire shot quickly

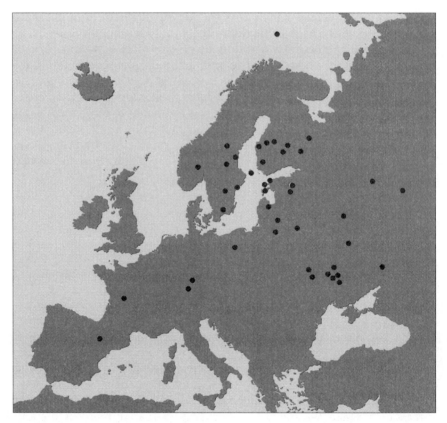

Significant European impact craters.

across the ship, injuring three seamen and breaking both gunwales, in the process ripping the planks from the stern of the starboard boat. The object bounced off the ship and exploded about 20m from the bow with a loud report, sparks flying from it like rockets. The captain reported no lightning or thunder at the time of the incident.

Research carried out at the Jet Propulsion Laboratory in Pasadena, California, by Kevin Yau, Paul Weissman and Donald Yeomans into ancient Chinese records dating from as far back as 700 BC reveals that there have been a large number of recorded meteorite falls and instances of meteorogenic human injury and fatality. Of more than 300 significant meteorite falls documented by Chinese scholars, there were seven events causing human fatality. In 1490, a remarkable downfall of meteorites struck the Ch'ing-yang region in Shansi Province with such ferocity that stones were said to have fallen like rain, and it was claimed that more than 10,000 people were killed. One of the later Chinese accounts (5 September 1907) details a large stone meteorite's impact on a house, which killed all of the unfortunate family members living inside.

Roofs of residential buildings all over the globe present to the sky one of mankind's most sensitive meteorite detectors. More than a dozen buildings across North America have been hit by meteorites over the last 30 years. On 30 November 1954, a 3.9kg stony-iron meteoroid smashed through the roof of a house in Sylacauga, Alabama, striking Mrs Hewlett Hodges on the hip and causing a contusion from which she later fully recovered. Mrs Hodges suffered from the twentieth century's only proven example of a meteorogenic injury.

Because of their proliferation, large surface area (especially in America) and predominantly outdoor habitat, motor cars represent a prime target for meteorite strikes. The first recorded meteoritic car smash happened in Benld, Illinois in September 1938, when a 1.8kg stone damaged a Pontiac Coupé. At Peekskill in New York State, at around 8pm on 9 October 1992, the trunk of Michelle Knapp's 1980 Chevy Malibu was smashed by a 12.2kg meteorite (an H-group chondrite) measuring around 12 x 14 x 11cm. The meteorite's entry into the skies above the northeastern US had been observed by thousands as it left a brilliant fiery trail which illuminated the sky, and it was seen from as far away as North Carolina. Later that evening, camcorder footage of the meteorite's blazing trail was featured on television news broadcasts. According to one report, a private meteorite collector offered Knapp the handsome sum of $69,000 for her rare car-smashing meteorite. The object is believed to be a smaller fragment of a much larger body whose pieces have not yet been recovered.

At midnight on 18 February 1995, a small stony meteorite entered the Earth's atmosphere in a steep southeastern trajectory over the Sea of Japan, brightened to a superb bolide which was observed by thousands, and ended its spectacular descent by pulverizing the boot of a white Honda car on the Japanese mainland. The meteorite was about size of a chicken's egg and weighed around 500g, and it was an example of the most common type of meteorite, an L6 chondrite. The object shattered into four individual pieces upon impact, and the largest fragment weighed 325g.

In modern times, less than a thousand meteorites have been seen to fall, whole intact specimens or fragments of which have later been recovered on the ground. According to Andrew Graham and his colleagues in the *Catalogue of Meteorites* (see Bibliography), the total number of authenticated meteorites in the world's collections amounts to 2,611.

SIGNIFICANT METEORITES

The largest single meteorite ever found is an iron weighing an estimated 60 tonnes, lying just where it fell at Hoba West, near Grootfontein in Namibia, many thousands of years ago. Clear traces of the considerable crater which it must have formed have long since been eroded; the

object, measuring 2.75 x 2.4m, lies exposed to the elements and is
slowly rusting away.

The Iron Tent

The somewhat chillier climes of Greenland were the setting for the
discovery of the Ahnighito meteorite, the second-largest iron meteorite
ever found. In 1894, the explorer (then Commander) Robert Peary was
guided by Inuit Indians (Eskimo) to the 31-tonne iron, along with two
smaller iron specimens nearby called the 'Woman' (3 tonnes) and the
'Dog' (400kg), as he was surveying an area near Cape York in northwest
Greenland. It transpired that for many years the Inuit had been merrily
chipping away at the objects – the largest of which they had named
Ahnighito (the 'Tent') – to obtain material with which to fabricate
knives and various other hunting implements. Generally speaking,
meteoritic iron is not ideal for such purposes, but this did not stop
Hollywood from claiming that the nineteenth-century American fron-
tiersman James Bowie fabricated his famous Bowie knife from a lump of
meteoritic iron far superior in quality to normal steel.

In 1895, the 'Woman' and the 'Dog' were shipped (although not
without great difficulty) south to the US. In 1897, driven in no small
way by the idea of further enhancing his public reputation – not to
mention an eventual payment of $40,000 – Peary and his team managed
to lash the Ahnighito meteorite to a large wooden pallet and transport
the giant iron across a substantial wooden bridge built between the
Hope, a 370-tonne, specially stabilized vessel, and the shore. Meanwhile,

Significant world impact events.

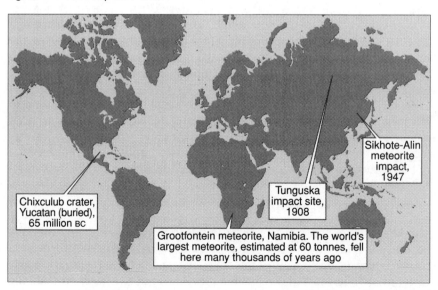

the Inuit received no compensation for the loss of their valuable natural metallic resource, and thereafter had little option but to buy their knifes ready-made from trading stations.

In truth, the loss of the meteorites was more a spiritual blow to the Inuit than a material loss, for they had been trading for better imported steel tools for many decades before their iron source was carted way. The Inuit considered the three meteorites to have been thrown from heaven by the a malevolent spirit called Tornarsuk, and to be distinct from run-of-the-mill local rocks, termed *weeaksue*. Because of this legend, Peary believed that the Inuit had actually witnessed the object's fall. This is highly unlikely, however, since the region had been colonized by east-ward-migrating Inuit only during the course of the last millennium, whereas the fall is believed to have taken place many thousands of years previously.

Before being installed in a museum, the Ahnighito meteorite lay for seven years on the New York dockside. Once museum space had been furnished and the meteorite's authenticity was fully verified, the massive object was paraded through the city streets like a live visitor from outer space, although the reception it received from the New York crowds was hardly as exuberant as that afforded John Glenn in the same streets in 1962. Today, the Ahnighito meteorite can be seen on display at the American Museum of Natural History in New York, where it still acts as a magnet for visitors from all over the world.

SIBERIAN SMASHES

Two significant cosmic impacts took place in Siberia in the first half of the twentieth century. In June 1908, a small cometary nucleus plunged into the Earth's atmosphere over Lake Baikal and exploded violently in mid-air, producing a massive shock wave that devastated a vast region of forest. This object is not considered to have been a typical meteoroid, and fragments of the impactor did not make craters in the ground. In February 1947, a large meteorite weighing around 70 tonnes exploded in mid-air over eastern Siberia, and many fragments impacted on the ground, producing numerous craters.

Cometary Impact at Tunguska

On 30 June 1908, a mighty cosmic impact was responsible for the de-struction of a vast area of Siberian forest in the basin of the Podkamen-naya Tunguska river. The 'Tunguska Event' was the biggest blow the Earth has received in recent history, and took the form of the nucleus of a small comet (or possibly a weakly bound meteorite) which exploded in the atmosphere at a height of around 6km.

The immense explosion, which centred over a sparsely populated area in central Siberia (60°55'N, 101°57'E), 400km north of Lake Baikal, was equivalent to 12.5 million tonnes (12.5 megatons) of TNT, or 2,000 times more powerful than the Hiroshima atomic bomb. Directly underneath the explosion, some 1,000sq km of forest and taiga were burned to a crisp; huge shock waves devastated an area of more than 2,000sq km and these were distinctly felt over 1,000km away. Thousands of reindeer and huge numbers of wildlife, plants and trees in the area were incinerated, but one extraordinary aspect of the incident was that only two people are known to have died as a direct consequence of the explosion.

Across the Tunguska region, many were alarmed at the tumultuous booming sounds of the comet's passage through the atmosphere and its violent terminal disintegration. The roar of the comet's death lasted for up to half an hour and was heard clearly within 800km of the explosion site. Most people attributed the cacophony to an extraordinary manifestation of thunder. Those who saw the comet's fiery atmospheric entry reported a blinding, broad red ball of light trailing a spear-shaped column of flame; after the final explosion, a brilliant flame split the morning sky and was replaced by a dark mushroom cloud of dust which rose to a height of 80km.

At the settlement of Vanovara, less than 100km south of the impact site, a villager named Semenov was subjected to the effects of the blast. In *The Tungus Event* (see Bibliography), Rupert Furneaux describes his plight:

> He [Semenov] had finished breakfast and had been sitting on the porch of his hut. He went into the yard and was just raising an axe to hoop a cask when suddenly 'the sky was split in two and its northern part was aflame'. He saw a great flash of light. A huge fireball filled the northern sky. He watched it for only a moment before he was forced to shield his eyes.... The heat became so intense that his shirt was almost burned off his back and his body was enveloped in flame... a mighty blast threw him from his feet to a distance of seven feet or more.... A terrible noise shook the whole house and nearly rocked it from its foundations. The glass and framing shattered and the ground split apart. He lost consciousness and his wife dragged him on to the porch. He revived to hear a noise like stones falling from the sky and rattling upon the roof.... The sky opened and from the north came a hot wind like the blast of a cannon.

Further north, closer to the explosion site, dozens of nomads and herdspeople were thrown into the air, and were bruised, concussed and shocked by the blast. On the southern edge of the zone of destruction, an old man named Lyuburman Shanyagir died of fright. About 30km southeast of the blast centre, another elderly man, named Vasily, was thrown 12m by the shock wave and hit a tree, breaking his arm and tragically dying soon afterwards. Thousands of reindeer belonging to four separate herds were also killed.

In Britain, a number of sensitive atmospheric pressure recorders called microbarographs simultaneously measured fluctuations in pressure which were distinct from most known atmospheric patterns. The instruments measured a series of rapid disturbances in atmospheric pressure on 30 June 1908, between 5am and 6am. The pressure fronts had been travelling at a speed later calculated to be between 308m and 323m per second. In comparison, similar pressure waves recorded in Britain at the time of the eruption of Krakatoa in August 1883 (modern history's most violent volcanic eruption) and propagated on the other side of the Earth, travelled at an estimated 314m per second. Comet expert, Fred Whipple, concluded that the energy responsible for producing these pressure waves was 5,000 times that which had produced the seismic waves related to the Tunguska event – in other words, the object had not hit the ground directly, but had exploded high in mid-air.

For years little was known about the event, owing to the extreme remoteness of the impact site and the country's turbulent political situation in the decade following 1908. It took the perseverance of Leonid Kulik (whose forename is of particular relevance to this book!), Curator of the Meteorite Department of the Mineralogical Museum of the Soviet Academy of Sciences in Leningrad, to conduct the first expeditions to investigate the site of the Tunguska event.

Kulik's attention had initially been drawn to the event in 1921, after reading a brief hand-written report which someone had copied from a Tomsk newspaper onto the back of an old calendar. The report claimed that the landing of a large meteorite in June 1908 had derailed a train at Filimonovo Junction on the Trans-Siberian railway, and that the meteorite lay buried in the ground next to the track.

Although the report was totally misleading, Kulik did not know this at the time; being a professional meteorite man, he was inspired to investigate further. The People's Commissary for Public Instruction provided Kulik with funds to visit the area; at Filimonovo, he was disappointed to discover that the alleged meteorite was just a natural Earth rock. But Kulik realized there was real substance to the story when he heard tales that a huge fireball had been seen over the Siberian skies in June 1908, passing over the eastern horizon and terminating somewhere north of the Stony Tunguska river with an incredible audible crash.

Kulik's expedition to the impact site, funded by the Society of Lovers of World Knowledge and the Soviet Academy of Sciences, went east from Leningrad in February 1927. The prime directive of the mission, issued by Academician Vernadski, was to locate the (suspected) meteorite as soon as possible and to make measurements of its size and composition. Vernadski reminded Kulik of the enormous scientific importance of the expedition, and stressed that positive results would possibly pay its costs a hundredfold.

After a gruelling two-month trek across 700km of harsh, freezing countryside, transporting his scientific apparatus and supplies on an assortment of pack-horses and reindeer-drawn sledges, Kulik and his team finally reached the outskirts of an area lined with row upon row of felled trees, all pushed over by the intense shock wave some 19 years previously. Of particular help to Kulik was a Tungus guide named Ilya Potapovich Petrov, who overcame his own fears that the region had been cursed by the local fire god Ogdy. Petrov was rewarded for his assistance by being paid two bags of flour, construction materials to rebuild the floor and roof of his home, and a quantity of cloth.

As the expedition progressed further into the zone of destruction and ascended to the summit of a hill called 'Sugar-Head', Kulik surveyed a scene of previously unimaginable devastation. The searing blast of the comet's explosion had uprooted, levelled and scorched an area of thick forest over a rough elliptical region measuring 60 x 40km. But no crater was found, and there were no appreciable meteoritic fragments.

Kulik visited the area again in May 1928, taking with him a movie camera to record the expedition. This unique chronicle, charting the trials and tribulations of a difficult expedition – including the moment when Kulik was nearly drowned in an icy river – was later edited into a short film called *In Search of the Tunguska Meteorite*. It was eventually concluded that the object causing the devastation had disintegrated in the Earth's atmosphere and had either completely vaporized or fragmented into a multitude of tiny pieces which landed over a large area. In spite of the scepticism of many of his colleagues, Kulik remained totally convinced that these fragments lay waiting to be discovered, and led a third expedition to the Tunguska region in February 1929.

Kulik hoped that unusual holes in the marsh represented pits excavated by the meteoritic fragments; he had tried to investigate these holes on his previous expeditions, but it had proved impossible to pump out the water from them adequately. This time, a team of workers dug a trench measuring 38 x 4m to draw the water away from one hole. The excavated soil was found to be devoid of meteoritic material, and a decayed tree trunk was discovered at the bottom of the hole, forcing Kulik to concede that the holes were not caused by secondary meteoritic impact.

There have been further investigations of the Tunguska site since Kulik's pioneering efforts. In 1958, half a century after the event, a scientific team made an extensive survey of the area. Since 1989, foreign scientists have been invited to join expeditions to the site of the great Siberian smash.

Because a Tunguska crater is absent, and no large fragments of the impactor have ever been unearthed, science has had little physical evidence into which to sink its speculative teeth. Soil analyses have revealed an abundance of silicate dust and tiny black, metallic, iridium-rich spheres less than a 0.1mm in diameter. There is a little evidence to suggest that the impactor was an E-type asteroid (stony in composition), with a diameter of more than 150m and weighing more than 7 million tonnes; this seems to be supported by the discovery of a very thin layer of dust with high levels of the element iridium, found at the 1908 level in core samples of the Antarctic ice sheet.

Pseudoscience has an acute sense of hearing; once the word 'mystery' is attached to any natural event without a glaringly obvious scientific explanation, crazy theories begin to appear. The Tunguska event has been explained variously in terms of a destructive parcel of antimatter popping up from another dimension; a black hole which sliced clean through the Earth; an early experiment in nuclear physics which got out of hand; and a faulty flying saucer which (to avoid harming human life) was deliberately crashed into sparsely populated Siberia by its considerate extraterrestrial crew.

The Sikhote-Alin Meteorite

On 12 February 1947, 5,000km east of Tunguska, the biggest meteorite of the century fell and burst over the mountains at Sikhote-Alin in the southeastern USSR (49°10'N, 130°40'E), north of the Chinese border. The objects blasted out 122 craters up to 26m wide and 6m deep. The total mass of the meteorites was initially estimated at 1,000 tonnes, but this figure was later revised and found to be in the region of 70 tonnes – still a big object. Some 27 tonnes of the impactor have so far been recovered from the site.

According to comet expert Martin Davidson (see Bibliography):

The [Sikhote-Alin] meteorite appeared at a height of 15 to 20 miles [24 to 32km] and was like a small, faintly luminous, reddish sphere. When it had dropped to a height of seven or eight miles, a trail of reddish-brown smoke was visible in its wake and could be seen more than a hundred miles away; the light of the meteorite was then described by an eyewitness as brighter than the Sun. At a height of about five miles [8km] it appeared to burst into dozens of fragments which fell almost vertically to Earth. The burst was accompanied by noises like thunderclaps and lasted four or five minutes; these noises must have been very intense as they were heard fifty miles [80km] away. The falling fragments produced craters scattered over an area of one-tenth of a square mile [25 hectares], and some of these craters were driven into hard rock. A preliminary analysis of its composition showed that it contained only six per cent of nickel, the remainder consisting chiefly of iron.

On 9 December 1997, the Internet buzzed with news of a meteorite impact of Sikhote-Alin proportions in Greenland. It was soon shown that although a meteorite had fallen, producing a brilliant fireball and a loud sonic boom as it fragmented, no major impact had in fact occurred.

5

COSMIC COLLISIONS

THE POWER OF COSMIC IMPACT

Since the Space Age began more than four decades ago, dozens of unmanned spaceprobes have revealed spectacular cratering evidence on every one of the solar system's solid worlds, including distant icy satellites and numerous rocky bodies much too tiny to have experienced any kind of conventional volcanic activity. Everywhere we have looked in the solar system, the signatures of cosmic impact are prominently carved in solid rock and ice, demonstrating that meteoroids and asteroids have been – and still are – a powerful planetary shaping force. The asteroids themselves display surfaces pockmarked with impact craters – as seen in recent dramatic spaceprobe images of the asteroids Gaspra, Ida and Mathilde, and in high-resolution Hubble Space Telescope views of Vesta.

In that strange bestselling classic of pseudo-science *Worlds in Collision* (1950), Immanuel Velikovsky claimed (among far more outlandish things) that the planet Venus had been formed after a giant comet had crashed into Mars. The few astronomers who considered the work worthy of criticism pointed out that there was no evidence to support such a dramatic scenario of interplanetary catastrophe. Today, astronomers still think Velikovsky's work has absolutely no scientific merit – and rightly so – but it is evident that views on interplanetary collisions have swung in favour of a historically more violent solar system in which major impact has been the chief sculpting force of the terrestrial planets, most planetary satellites and the asteroids. Most of the major impact craters visible on the Moon, Mercury and Mars were excavated more than two billion years ago as these bodies swept up debris left over from the solar system's birth.

CRATERS OF THE MOON

To the naked eye, the Moon appears as a disc 0.5 degree wide, covered with a patchwork of bright and dark areas. The dusky features that compose the 'Man-in-the-Moon' are vast, ancient asteroidal impact scars that were filled with flows of dark-coloured lava soon after their formation. Those with keen eyesight can discern several bright spots on the lunar disc marking the sites of the impact craters Tycho, Copernicus, Kepler and Aristarchus, each surrounded by its own systems of rays – material of the Moon's crust that was flung out far and wide by the tremendous energies released in the impact process. Whenever you gaze at the Moon, remember that much of what you can see is the result of a catastrophic period of asteroidal and meteoroidal bombardment that befell our satellite at around the same time that the first primitive terrestrial life-forms were evolving.

Through a telescope, the face of the Moon, Earth's only natural satellite, resolves into a chaotic mixture of undulating grey lava plains and majestic highlands pockmarked with countless thousands of craters. It has been estimated that the lunar surface contains at least 3 billion craters larger than 1m across, but only craters larger than 1km can be resolved in ordinary backyard telescopes. Ever since Galileo wondered at the Moon's impressive terrain through his tiny handmade telescope at the beginning of the seventeenth century, astronomers have attempted to explain how the lunar craters were formed.

Early Theories

Early this century, the scientist D. P. Beard suggested that the craters might be the circular remnants of vast coral atolls formed in ancient, shallow lunar oceans. In 1925, E. O. Fountain envisioned the craters as round depressions left in the Moon's surface after pockets of ice had melted; interestingly, this mode of planetary sculpting has recently been cited as the cause of many of Mars' unique valleys and depressions. Having seen the horrifying destruction unleashed by the atomic bomb, in 1949 one Sixto Ocampo, a Spanish engineer, painted the way-out picture of a terrible lunar war that had obliterated the lunarian combatants and left their home world pitted with nuclear explosion craters. Many other weird and wonderful models of crater formation – including recent claims that some of the Moon's surface has been altered by intelligently orchestrated (alien) excavation – may all safely be discounted as nothing more than poorly researched science-fiction tales.

Some of the first practical experiments in planetary geology were conducted as long ago as 1667, when the English scientist Robert Hooke boiled pots of powdered alabaster and water to simulate a hypothetical ancient, hot lunar surface. Hooke noted the changing appearance of the porridge-like mixture as it frothed, bubbled and splattered into action.

Once the experiment had cooled and consolidated, the scientist saw seemingly obvious similarities with the Moon's surface; he concluded that the real lunar craters might be the solidified rims of gigantic bubbles that had once burst forth from the hot lunar rock.

Excited at the prospect of recreating the first chapters in the life of Earth's only natural satellite, Hooke also lobbed small pellets into mixtures of clay and water to produce features which resembled impact craters surrounded by rays. However, the scientist ultimately rejected the impact theory of crater formation because the existence of large meteoroids in space capable of wreaking such devastation was then unknown, and probably unthinkable. Indeed, Hooke poured scorn upon the very notion that lunar craters were formed by impacting meteoroids and asteroids, saying that 'it would be difficult to imagine whence these bodies should come'.

Long before Hooke's era, philosophers had conjured up all manner of theories explaining meteorites in terms other than that they were objects from outer space. From the time of Aristotle, who wrote about various celestial phenomena in his *Meteorologica* (*c*.340 BC), to the beginning of the eighteenth century, it was generally believed that meteorites were stones formed in the Earth's atmosphere in a region below the lunar 'sphere'. Yet many scholars seriously doubted that meteorites fell to Earth under any circumstances whatever, whether they came from the Earth's atmosphere or beyond.

Scientific investigations soon began to incorporate discoveries like electricity into their meteoritic theories, leading to claims that terrestrial vapours and gases were ignited by unseen electric discharges to produce meteors and fireballs. This was seemingly supported by claims (now known to be erroneous) made by the French scientist Pierre Bertholon, who said that meteors and fireballs, along with their explosions, could actually be reproduced within the confines of laboratories using large static-electricity generators.

Nineteenth-Century Science

It took the efforts of the German physicist Ernst Chladni (1756–1827) to bring the theory that meteorites have extraterrestrial origins well and truly into the scientific limelight. In 1794, Chladni published an intricate analysis of meteors, fireballs and meteorites, and asserted that there was no evidence that iron meteorites could have been formed by any known terrestrial process. Furthermore, he suggested that the debris responsible for meteors and fireballs was probably a by-product of the formation of the solar system, or alternatively that meteoritic material represented fragments of a planet that had once exploded. The latter idea firmly took root when the asteroids began to be discovered orbiting between Mars and Jupiter, starting with Giuseppe Piazzi's discovery of Ceres, a minor planet measuring 920km in diameter, on 1 January

1801 – the very first day of the nineteenth century. After Chladni's painstaking investigations and daring speculation, science found that it could no longer continue to bury its head in sheer disbelief. The planet Earth, once thought to be the inviolable hub of the universe, began to be seen for what it truly is – a small planetary outpost vulnerable to natural bombardment from the depths of space.

With the tonic of Chladni's research beginning to circulate in science's anaemic blood stream, a single event at the beginning of the nineteenth century would dramatically confirm ideas of cosmic impact. On 26 April 1803, a large meteoroid entered the atmosphere above France and exploded violently. More than 3,000 substantial fragments of this size-able rock from outer space were deposited over the countryside near the village of l'Aigle, across an elliptical area measuring some 9 x 4km. The subsequent in-depth scientific enquiry, headed by the scientist Jean-Baptise Biot, persuaded many scientists and philosophers of the undeni-able reality of stones falling to Earth from interplanetary space.

Hundreds of local citizens had seen the l'Aigle fall, most of whom were credible and coherent witnesses whose evidence it was impossible to dismiss. The objects found scattered about the countryside did not resemble the local rock, and Biot concluded that the evidence he had heard and found, overwhelmingly favoured the landing of a quantity of extraterrestrial stones.

As science finally began to accept the astonishing fact that Earth is occasionally bombarded with substantial pieces of solid material from outer space, the astronomer Franz von Gruithuisen made an earnest attempt to promote the impact theory of lunar crater formation. The astronomer's ideas, first postulated in 1829, were sound enough, yet they tended to be regarded with amusement by many of his fellow scientists. Gruithuisen was a proficient astronomical observer, but his character was somewhat eccentric. After his announcement, in 1822, that he had discovered evidence of an impressive lunar city located near the centre of the Moon's disc, Gruithuisen's credibility among his peers had taken a dramatic tumble. Not surprisingly, the astronomer's more sensible later claims that the Moon possessed a visibly bombarded sur-face fell largely upon deaf ears.

Mainstream science in this era maintained that the only plausible way in which the Moon could have become so cratered was through volcanic activity. The idea that meteoroidal missiles could have shaped the Moon's face in any major way was as untenable as Gruithuisen's fabulous lunar city. The lunar impact hypothesis took a long time to percolate through the thick strata of scientific thinking, attaining near-universal support – substantiated by concrete evidence gathered on the Moon's surface – only during the latter half of the twentieth century.

At the end of the nineteenth century, the first credible investigations into lunar impacts and the craters likely to be produced by meteoroids and asteroids were conducted by the American geologist Grove Gilbert. He ultimately discounted long-cherished ideas that the craters were grand volcanic vents, and put forward his theories on impact after making systematic telescopic observations and conducting painstaking laboratory experiments. Gilbert's fellow countryman and scientist, the geologist Josiah Spurr, saw little to commend in Gilbert's work. Spurr was an ardent supporter of lunar vulcanism who would not allow the lunar collision of the tiniest meteoroid to enter his somewhat closed mind. Unlike Gilbert, Spurr rarely observed through a telescope, choosing instead to base much of his work upon perusing photographs of the Moon's surface. In his *Geology Applied to Selenology* (1944) he wrote:

> I have encountered nothing that would lead me to interpret any [lunar] feature as the result of meteoric infall.... The sculpture of Moon features in general is quite sharp. Indubitable eruptive craters... are unmarked by evidence of any later catastrophe, such as pelting by celestial missiles.

It is ironic to think that the small lunar crater in the Marsh of Decay named in Spurr's honour is actually a meteoroidal impact feature which was later modified by vulcanism and flooded with lava!

Missions to the Moon

Through the manned Apollo missions, which explored and sampled six geologically diverse lunar locations over a total area of tens of square kilometres, not to mention dozens of sophisticated unmanned lunar probes, an enormous amount of hard information about the nature of the Moon's surface has been gathered. Several international lunar science teams were given the opportunity to analyse the data obtained on the lunar surface and to make minute studies of the lunar rock and soil samples.

To summarize, Apollo found overwhelming evidence to support the theory that most of the Moon's craters are meteoritic impact features; even in those areas where vulcanism had once been strongly suspected before the manned missions, little evidence for volcanic crater-forming eventually became known. The astronauts also set up small science stations at each landing site, four of which incorporated sensitive seismometers which could detect tremors in the Moon's crust. Some of the vibrations recorded were due to deep Moon-quakes, but many were found to have been caused by the impact of meteoroids on the Moon's surface. After careful analysis of the signals, it was possible to estimate roughly where each meteoroid had landed and the force of impact. In a 2.5-year period commencing in 1973, most of the 815 recorded meteoroid impacts were randomly distributed. But on three occasions – November and December 1974 and June 1975 – the Moon seems to have experienced a barrage of large meteoroid strikes as it passed through unseen meteoroid swarms.

Meteoroids are the principal cause of erosion on the lunar surface. No meteor flashes ever become visible in the skies of our airless sister planet, since there is no friction to heat them as they speed Moonwards, and every meteoroid slams into the lunar surface at its original interplanetary velocity. Like sandblasting, millions of meteoroids and micrometeoroids continually scour the lunar surface at speeds of tens of thousands of kilometres per hour. Each tiny hit is capable of pulverizing or vaporizing a small portion of the surface, wearing down solid rock and producing the fine lunar soil.

In November 1966, shortly after that year's spectacular Leonid meteor storm, the Cornell University Centre for Radiophysics and Space Research in New York State announced that their studies of the photographs returned by the US Surveyor 1 and the Russian Luna 9 soft-landing lunar probes had led them to conclude that the Moon's surface is covered with at least 150mm of fine soil. Using dynamite explosions in rock powder to simulate the impact of meteoroids, a type of soil was created which had the same properties as lunar soil when a model of the Surveyor footpad was dropped into it. Moreover, the soil was deemed perfectly capable of supporting heavy spacecraft and manned activity. Some of the rocks imaged in the Surveyor and Luna photographs were thought to be delicate conglomerates of soil. Researchers claimed actually to have created objects that looked exactly like solid rocks, but were so fragile that they fell apart once touched.

The lunar soil was eventually found to have the same consistency as wet sand, and the astronauts who kicked up Moon dust found that it clung persistently to their spacesuits. Astronaut Al Bean of Apollo 12 was to make good use of dusty bits of his spacesuit back on Earth by grinding them up and incorporating the residue into many of his space paintings. His artworks have a unique selling point – they are the only paintings of the Moon made with real lunar pigments.

Countless high-speed meteoroidal impacts over billions of years have created the Moon's fine soil layer. Just as fine sand is formed on the seashore by the relentless pounding of the waves against apparently indestructible cliffs, the Moon's rocky crust has been exposed to a perpetual sandblasting from space. Meteoroid impacts are the Moon's main cause of erosion. The footprints left by the astronauts are already beginning to be worn away by this erosion, but the process is so slow that (unless large impacts wipe them out) they will remain visible for many thousands of years to come.

Twelfth-Century Lunar Impact?

No really big impact has shaken the Moon's surface for millions of years. An ancient twelfth-century manuscript written by the English monk and chronicler Gervase of Canterbury (*c.*1141–*c.*1210) is thought by some to represent an account of a medium-sized meteoroidal impact on the Moon's far side, which is said to have taken place in June 1178.

A passage in the chronicle of Gervase, now preserved in the library of Trinity College, Cambridge, relates to a spectacular lunar observation which was made by five men in Canterbury, Kent, who happened to be admiring the narrow crescent Moon one evening. The men are said to have been surprised to see 'a flaming torch' spring up from the upper limb of the crescent, 'spewing out, over a considerable distance, fire, hot coals and sparks'.

Immediately afterwards, Gervase reported the Moon to have 'writhed like a wounded snake' and assumed a blackish appearance all over. It is possible – though hardly certain – that these people observed the effects of the impact of a meteoroid just past the Moon's northeastern edge.

The first attempt to pin down the crater likely to have been excavated by the collision was made by Jack Hartung of New York University, who studied photographs of the far-side area in question. Situated at 36°N, 103°E lunar co-ordinates, well past the northeast lunar limb, a bright crater named Giordano Bruno sits at the centre of an impressive system of bright rays. Bruno, 20km across, is a sharp feature that looks very young by lunar standards. If those men of Canterbury had witnessed a large meteoroidal impact, then, more than any other crater in the area, prominent-rayed Bruno is likely to have been the result. Bruno's spectacular ray system extends on to the near-side of the Moon, and some of the dark Mare Crisium is faintly streaked with this light-coloured debris. If the crater had been formed on the near-side of the Moon, then it would easily be visible with the naked eye, a bright spot ranking alongside the stunning ray systems of craters Copernicus and Tycho (aged 900 and 100 million years respectively).

Mechanics of Lunar Impact

The meteoroid responsible for the Bruno event postulated above carved a crater out of solid rock which is 20km wide, releasing energies which flung out vast amounts of material to distances of hundreds of kilometres from the impact site. The impacting object may have been as small as 1km in diameter. The size and shape of impact craters depend on the composition of the meteoroid, its size, mass, velocity and direction, as well as the type of rock it impacts upon.

A small, fragile cometary nucleus dashing itself on the Moon's rocks will hardly make a dent. It is worth mentioning that delicate whorls of light-coloured material have been discovered on the lunar surface, notably the feature called Reiner Gamma in the Ocean of Storms (Oceanus Procellarum) and several similar markings on the far-side. These wispy formations are probably areas marking the impacts of fragile cometary nuclei, which are nothing more than large balls of dirty ice without much integrity other than their weak gravitational bonding.

A meteoroidal body of the same mass as a small comet will be much stronger and gouge out a substantial crater in the lunar crust.

For a low-speed meteoroidal impact – say, a body with a velocity of just 17,000km per hour – the object will have a kinetic (speed) energy equal to its own mass in TNT. These types of collision would produce a simple mechanical impact crater; the meteoroid is shattered on impact and the fragments lie buried beneath the lunar surface at the site and distributed around the local countryside, mixed with excavated material in an ejecta system. Impacts of Moon probes and discarded rocket stages directed at our satellite have blasted out quite hefty craters, some of which have been prominent enough to be photographed from lunar orbit.

The angle of approach and direction from which the meteoroid came would decide the crater's eventual shape. If the low-speed impact had been at a very oblique angle and from the west, then an elliptical crater with an ejecta system spreading to the east would be formed.

A bigger, faster meteoroid weighing a million tonnes, travelling through space at a speed of 30km per second – equal to the Earth's speed around the Sun – will have a kinetic energy of 10 million million million joules.

With such a fantastic amount of energy – equivalent to more than a billion tonnes of TNT – the fast-moving meteoroid would slice into the Moon's solid crust like a hot knife through butter and would burrow down into the Moon to distances several times its own diameter before stopping. Its kinetic energy would be converted into other forms, such as the mechanical energy of shock and fracture, heat energy and seismic waves, much of it being transferred directly into the local lunar rocks.

If the meteoroid was stony, a temperature of several million degrees would be produced, quite sufficient to vaporize the original meteoroid utterly. An ultra-hot bubble of gas would scorch the rocks around it to temperatures of a few hundred thousand degrees, but the sheer weight of the lunar rock above the gas bubble would not be enough to trap it for more than a fraction of a second. The local rocks would rapidly heat and expand, resulting in an explosion which would blast out a region many times the diameter of the original missile. The angle at which the meteoroid originally approached the Moon would not have much relevance to the resulting crater shape, because the energy released would be akin to a point detonation, forming a near-circular depression.

In a few cases, such an impact will be powerful enough to blast some material off the Moon's surface, so that it is ejected into space and escapes the Moon's gravity; some of this material may make its way to the Earth as meteorites. Lunar meteorites are one of the rarest rock types on Earth, with only 13 known examples. All but one of these bits of the Moon have been discovered on the south polar ice sheet by the Antarctic Search for Meteorites (ANSMET). By taking advantage of the pristine, snowy white environment to spot any unusual debris lying upon it, ANSMET has made many significant meteorite finds over the past couple of decades. The majority of Antarctic finds have been regular iron, stony-iron or stony type meteorites, but some have undoubtedly come from the surface of the Moon, Mars and other solid bodies across the solar system.

Luna's Battered Face

To the naked eye, the mottled face of the Moon shows obvious signs of impact. The circular patches occupied by the lunar seas are vast impact 'craters', formed by asteroidal collision and filled with dark lava flows early in its history. The bright highlands are full of craters, and the ray systems surrounding four impact craters – Tycho, Copernicus, Kepler and Aristarchus – can be seen by the keen-sighted the under the right conditions.

About 4.3 billion years ago our hot, molten, embryonic satellite began to develop its first crust, pieces of which are still in existence. Several extremely ancient impacts on this crust – producing formations now so eroded that they can hardly be seen – are believed to have taken place around this time, including the vast far-side 'Aitken' basin, the near-side 'Gargantuan' basin and the 'Schiller Annular Plain' near the southwestern limb (these are all unofficial names). These very old scars were followed by the asteroidal impacts which produced the basins of Nectaris (3.92 billion years), Serenitatis (3.87 billion), Crisium and Humorum (both 3.8 billion years). No less than 50 major lunar impacts are believed to have taken place between 4.2 and 3.8 billion years ago.

Two events of similar magnitude – the impacts of the asteroids that carved out the Imbrium and Orientale basins at 3.85 and 3.8 billion years ago respectively – mark the end of the first era of major lunar bombardment. Both features were formed by asteroids less than 100km in diameter. Although broad fronts of lava spread across the fresh Imbrian basin, totally burying the inner structures, Orientale's flooding (and the filling of other far-side features) was restricted to its centre and just a few outlying areas. Orientale is therefore one of the solar system's best preserved major impact sites. The eastern maria of Tranquillitatis, Crisium and Fecunditatis were formed around 3.5 billion years ago, preceding the flooding of Imbrium and Procellarum in the western hemisphere by several hundred million years.

More than a third of the near-side of the Moon was covered with lava flows, compared with just a minute fraction of the far-side. Only a few far-side impact basins were flooded with lava, because the crust overlying the hot melt layer was significantly thicker and it experienced fewer really big impacts than the near-side. Large-scale lunar vulcanism had virtually ceased by around 2.5 billion years ago, accompanied by a decline in the rates of major impacts.

Around 2.4 billion years ago, the impact of a large meteoroid formed Eratosthenes, a crater whose ray system has faded and is not now obvious. Other major impact craters whose ray systems can be identified with the naked eye include the prominent craters Copernicus, Aristarchus and Tycho, which were formed around 900, 300 and 100 million years ago respectively. The current period of lunar history has seen little in the way of major impact, but relentless surface erosion by countless meteoroids has rounded the once-jagged lunar hills and produced the layer of fine soil called the regolith.

The latest period in lunar history began in September 1959 with another impact crater, formed by the crash-landing of the Russian space-probe Luna 3. Since then, the impacts of more than 28 spaceprobes and bits of rocket have blasted their own craters here and there on both the near-side and far-side of the Moon. The future activities of mankind may ultimately transform the appearance of the Moon so much that the excavations will be noticeable through the eyepiece of a small telescope on the Earth.

TARGET: EARTH

Scholars of old imagined that our planet lay firmly at the centre of the universe – a solid, unmovable raft upon which a supreme creator had allowed mankind to flourish without any natural outside interference. The Earth was not perfect, for natural disasters of one kind or another occasionally wreaked a fair degree of havoc, upsetting the terrestrial inhabitants. But at least natural dangers were thought to be produced by forces within or near to the Earth itself, whether it was an internally fuelled volcanic eruption or the mighty atmospheric upheaval of a hurricane. It once seemed impossible (even heretical to suggest) that the tranquil heavens above could ever host forces capable of damaging the Earth or its life.

That view changed 200 years ago, when science began to accept the notion that the Earth was not inviolable by extraterrestrial forces. Scientists have proved that outer space is populated by meteoroids and asteroids which occasionally slice through the Earth's atmosphere and do damage. Throughout its 4.6 billion year history, the Earth has been on the receiving end of countless meteoroidal and asteroidal impacts. Because our planet is such a large target, with an area 14 times that of the Moon, the bombardment in Earth's early history must have been far more intense than that of our sister planet. It has been estimated that in the past half-billion years alone the Earth has been subjected to the impacts of around 2,000 asteroid-sized bodies. Before this era, the cosmic catastrophes which befell our planet were undoubtedly more frequent, if the face of the Moon tells us anything about impact rates.

Maps of the Earth's surface do not make it obvious that impact has been a great sculpting force. Most of the evidence for ancient bombardment has been utterly erased by the powerful forces of plate tectonics and vulcanism, in conjunction with the dynamic atmosphere, hydrosphere (oceans, rivers and lakes) and biosphere (plant, animal and human life).

Mechanics of Earth Impact

Most cosmic bodies that encounter the Earth's atmosphere are travelling at relative velocities ranging from 11km to 73km per second. Meteorites with a mass of up to 100 tonnes are slowed significantly by the Earth's atmosphere, but they can hit the Earth with enough force to penetrate the ground, boring a pit and eventually coming to a halt some distance beneath the Earth's surface. The impactor (or pieces of it), although it has been intensely heated by friction with the atmosphere and the ground, will survive as a meteorite.

Larger impactors, with enough momentum to slice through the atmosphere and retain a greater proportion of their original interplanetary speeds, will explode violently on striking the Earth. At 17,000km per hour (5km per second), an impactor has a kinetic (speed) energy equal to its own mass in TNT. Since kinetic energy increases with the square of velocity, an object of the same size but travelling ten times faster (50km per second, or 170,000km per hour) will have not ten but 100 times the explosive potential (10 x 10) of its own weight in TNT. A 10,000 tonne meteoroid impacting on the Earth at such a speed will explode beneath the Earth's crust with the force of a million tonnes of TNT.

The smallest impact craters are pits excavated from the ground and their shape is dependent on the angle at which the impactor hit and the strength and composition of the material at the impact site. A small meteoroid impacting from a steep angle will produce a simple bowl-shaped depression surrounded on all sides by ejected debris. The same object hitting the Earth from a shallow angle will carve out a crater that is long and pear-shaped as seen from above, with a pile of debris flung out in one direction from the 'blunt end' preceding the impactor. The ejected debris can, in turn, produce its own secondary cratering, giving rise to a crater chain of sorts.

Major explosive impacts produce large craters which, because they have been derived from what is effectively a point explosion deep beneath the surface, are circular in outline. The impact can set up concentric waves within the crust, which later fracture and produce multiple ring structures with slumped, terraced walls. Large, young impact craters usually have some sort of central elevation, typically a range of hills or mountain blocks, which are produced by the immediate elastic rebound of the local rocks after being compressed during impact. The shape of the blast wave propagated beneath the crust may also leave a central elevation directly beneath the point of explosion. However, this feature is never higher from the floor than the rim of the crater itself. It has been demonstrated that central elevations occur more readily in impact explosions which have happened at shallow levels beneath planetary crusts.

The rocks around the impact site will be shattered and metamorphosed by the intense pressure and heat generated during the event. Two rare high-density, high-pressure varieties of the mineral quartz, called Coesite and Stishovite, are seen as strong indicators that a meteoroidal or asteroidal impact has transformed rocks. The minerals were both detected at the Barringer Crater (see page 94) and have since been found within rocks at the sites of other suspected impacts.

Earth's Mobile Crust

The Earth's surface is such a dynamic place that even the largest impact features may be eroded, deformed and buried within a short period in the geological timescale. Plate tectonics are a geophysical mechanism, fuelled by convection forces within the Earth's hot, fluid mantle layer beneath the solid crust, which has moved (and is still moving) entire continental and oceanic plates around the globe. Areas where plates meet and interact can be constructive or destructive. When continents collide, the edges of the plates crumple and are thrust up to form whole mountain ranges – like the Himalayas, where the subcontinent of India has collided with Asia, piling and folding vast sheets of the Earth's crust into one of the solar system's most impressive highland plateaux.

There are places where oceanic plates encounter continents and, to put it simply, are compelled to dive underneath them. Such 'subduction' zones lie around the edge of the Pacific Ocean in the 'ring of fire', so-named because of the volcanic activity which takes place as the subducted crust heats, intrudes into the cooler crust overhead and creates volcanoes. And new plate material is being extruded in extensive oceanic mountain ranges, most notably along the Mid-Atlantic Ridge, a place where fresh lava is erupting along the margins of the North American-Eurasian plates and the South American-African plates. The overall picture of relentless constructive and destructive plate activity at a relatively fast rate, overlain by powerful erosion forces, means that ancient impact scars are difficult to identify with any certainty.

Earth's development is thought to have been influenced greatly by ancient impacts. If we had escaped major cosmic assaults, our global maps would appear completely different from those with which we are familiar. For example, an impact of the same magnitude as that which excavated the Moon's Imbrian basin some 3.85 billion years ago would have smashed through the young Earth's thin crust easily, introducing a vast system of faults centred around the point of impact and stimulating widespread vulcanism. Such an event is likely to have triggered an increased phase of plate tectonics and changed the course of continents.

The ultimate configuration of the Earth's plates would have been noticeably different as a result of just this one Imbrium-sized impact.

Modern science has very little knowledge of the Earth as it was 4
billion years ago, nor are we ever likely to discover details of how the
face of the Earth was sculpted by external forces in its early history. The
oldest rocks ever found on the Earth date from 3.8 billion years ago,
after the time when all of the Moon's big basins had been formed. When
you look at the Moon, you can see features which have been preserved
more or less intact from a time long before the sketchy knowledge of our
own planet's history begins.

If you study a map of the Earth, it is not difficult to identify vague
outlines which seem to suggest the presence of eroded impact craters.
The largest and most prominent of these lies on the eastern shore of
Hudson Bay in Canada, which some believe to be the semicircular
outline of an ancient crater 450km, across whose central mountains are
clearly marked by the Belcher Islands offshore. But many other apparent
impact features – bays and coves which pose as the rims of old meteoroidal impacts – have been formed solely by terrestrial processes. The
Gulfs of Mexico and St Lawrence in Canada, the Bahia Samborombon
and the Golfo San Jorge in South America, the Sea of Japan and Antarctica's Weddell Sea have all been offered as candidates for ancient impact
scars, but these designations have not been supported by the hard geological evidence, which all points to purely terrestrial causes.

The impact of an asteroid just 10km across – small by cosmic standards – would induce movements in the Earth's crust like waves higher
than houses propagating away from the epicentre. Accompanying this, a
blast of air heated to 500°C and moving at a speed of 2,500km per hour
would destroy anything living up to distances of 2,000km. The impact
of such a body on the ocean would create massive tidal waves which
would swamp the world's coastlines, to devastating effect.

A HISTORY OF TERRESTRIAL IMPACTS

Asteroidal Impact and the Death of the Dinosaurs

Around 65 million years ago, a large asteroid burned in the atmosphere
over the Caribbean and impacted on the (now) Yucatan peninsula. The
dinosaurs of the late Cretaceous period who saw the blazing asteroid, felt
the mighty earthquakes which followed the impact and lived to roar
their discontent, could not have realized that the event would mark the
beginning of their demise and extinction.

The theory that many dinosaur species (along with the pterosaurs,
ichthyosaurs, plesiosaurs and mosasaurs) died out because of the effects
of an asteroidal impact had been proposed long before any evidence for
an end-of-Cretaceous crater had been discovered. The theory suggests
that the enormous quantities of dust thrown up into the high atmos-

phere were sufficient to reduce significantly the amount of solar heat reaching the Earth's surface, hence killing off the algae, vegetation and small animals on which the dinosaurs depended for food.

In 1978, traces of a huge 200km impact crater buried hundreds of metres underground were detected in a series of aerial magnetic surveys conducted by the Mexican oil company Pemex. Named Chicxulub Crater after a town in Mexico which overlays the site, the formation's outline is comparable in size with the lunar crater Clavius; it is believed to be just the right age and size and in the right location to be the scar of the asteroid responsible for the dinosaur catastrophe.

Somewhat ironically, an asteroidal impact around 38 million years ago on the northern side of the Gulf of Mexico may have been instrumental in creating a vast natural haven for wildlife – the Everglades marsh in Florida. Conventional geological theory postulates that the limestone basin overlain by the swamp (totalling nearly 13,000sq km) has been formed over millennia by erosion by waters overflowing Lake Okeechobee in the north. But Edward Petuch of Florida International University has suggested that the limestone basin represents a crater carved out by an asteroidal impact.

Geological surveys have shown that the limestone bedrock underlying the Everglades is crossed with a system of rock fractures reminiscent of a major impact. Coral reefs soon formed on the raised rim of the impact crater, building up into a large circular atoll. The Earth's climate at this junction of geological time (between the Eocene and Oligocene eras) changed, leading to a generalized cooling. At the same time, sea levels rose as the world's land masses increased in size.

As a result, the coral tended to migrate northwards in search of a warmer environment with shallower waters, a process which extended the atoll in the north–south direction. The coral walls of the enlarged island atoll eventually enclosed the waters completely to form an inland lake. Once sedimentation began to fill the feature and the original salt water had been replaced by fresh water, savannah grasses and virgin forest advanced inwards and reduced the size of the lake. Today, the 1,800sq km of Lake Okeechobee is all that remains of the original lake.

One sure indicator that a meteoroid or asteroid has caused a particular crater is to find pieces of the original impactor buried beneath the crater or distributed around the local countryside. So far, only a dozen craters around the Earth are known to be associated with meteoritic fragments. Such indubitable meteoritic features are only the relatively young ones in the cosmic timescale, because small craters much older than 100,000 years are liable to have been eroded beyond immediate recognition. The tell-tale pieces of meteoritic material lying around the impact site, especially iron fragments, weather rapidly to rust and change in appearance, blending into the country rock so well that only an expert eye may be able to identify them.

Mr Barringer's Elusive Meteorite

The famous Barringer Crater in Arizona is undoubtedly the best known of all terrestrial impact craters, and is the largest of the world's meteoritic craters known to possess fragments of the original meteoritic impactor. The Barringer Crater is the first ancient impact crater to have been discovered in modern times, and was stumbled across in the 1870s by westward-moving pioneers on the hunting grounds of the Apache Indians. The feature is sometimes referred to as the 'Meteor' Crater – a misnomer, since a meteor is merely the flash of light visible in the sky when a small meteoroid burns up in the atmosphere.

The huge pit, 1.3km in diameter and 175m deep, with a rim raised over 40m above the surrounding terrain, was soon recognized for what it was – a giant hole in the Earth's crust excavated by the impact of a large meteoroid, which exploded with the energy of 20 million tonnes of TNT. The crater is not large in comparison with those seen on the Moon, and if transplanted there it would be classed as just a minor craterlet and would require quite a large telescope to be discerned clearly from the Earth.

In 1891, the crater was visited by the Philadelphia geologist Dr Albert Foote, who was commissioned to survey the site by the Santa Fe Railroad Company to determine whether or not there was anything of mineral value at the site. Foote came across scores of meteoritic iron fragments around the local area, and in his excitement went so far as to postulate that coarse diamonds might exist within the (presumed) buried meteoritic body. Foote's vision of a large and commercially exploitable nickel-iron/titanium-rich mass lying buried beneath the crater prompted the mining engineer Daniel Barringer to conduct penetrative investigations, beginning in 1904.

Two decades later, at a total cost in excess of $600,000 ($120,000 of which was Barringer's own money), more than 100 bore-holes had been sunk deep into the rock around the site, but sadly there was no evidence to suggest the presence of an appreciable body underground and the project was consequently abandoned. In 1929, after Forest Moulton (an eminent mathematician) had determined that any meteoritic iron mass buried beneath the crater, if indeed it existed at all, was probably too small to be of any commercial value, Barringer suffered a stroke and died.

Nowadays, we realize that it is erroneous to assume that large impacting bodies, like that which created the Barringer Crater, are just dumped into the Earth like a stone thrown into sand. With their considerable mass and high velocity, big, fast meteorites penetrate deep into the crust and heat the rock about them to very high temperatures, to form a high-energy bubble which explodes and redistributes the meteoritic material and melted local rock over the surrounding environment.

The Barringer Crater is currently estimated to have been formed

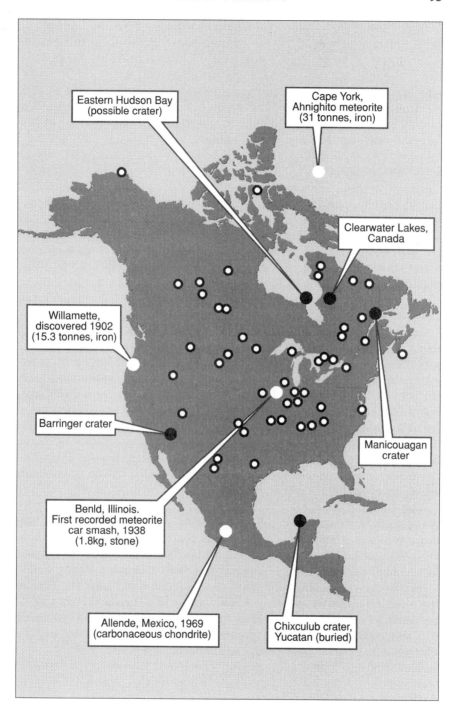

Eastern Hudson Bay
(possible crater)

Cape York,
Ahnighito meteorite
(31 tonnes, iron)

Clearwater Lakes,
Canada

Willamette,
discovered 1902
(15.3 tonnes, iron)

Barringer crater

Manicouagan
crater

Benld, Illinois.
First recorded meteorite
car smash, 1938
(1.8kg, stone)

Allende, Mexico, 1969
(carbonaceous chondrite)

Chixculub crater,
Yucatan (buried)

Significant North American impact craters and meteorite falls.

around 49,000 years ago by an iron-nickel body some 70m across and weighing as much as 2 million tonnes. Most of the local inhabitants who witnessed the meteoroid's catastrophic arrival probably did not live to tell the tale, but our human ancestors all around the world undoubtedly experienced the effects of the impact to some degree, either by feeling the Earth tremors shake the ground beneath them, by sensing the mighty pressure wave or by seeing atmospheric effects produced by the large cloud of dust released into the atmosphere.

Canadian Craters

Canada's rugged landscape has been altered by millions of years of glacial activity, and the western region has been modified by plate tectonics. Since the late 1950s, Canadian geologists have been particularly active in identifying ancient impact craters strewn about their vast country. This has been made possible by good aerial photography, including images obtained by satellites, and also the tell-tale geological signs of the presence of material of a different composition having been brought to the surface from beneath the crust after impact. By 1997, some 26 probable impact craters and 12 possibles had been identified.

The largest Canadian astrobleme is a 450km-diameter crater marking the eastern shore of Hudson Bay, a feature which would put all of the Moon's near-side walled plains in the shade. One of Canada's most interesting impact craters lies to the east of the rim of this feature and takes the form of two craters next to each other, Clearwater West and Clearwater East, with diameters of 32km and 22km respectively. They were formed simultaneously by a double meteoroid which impacted around 300 million years ago. An interesting size comparison can be made between the Clearwater craters and the lunar craters Clavius D and C, the two central craters in the arc running across Clavius' floor and both visible through a small telescope.

The 140km-diameter Sudbury basin, north of Lake Huron in Ontario, is the mark of an asteroidal collision which occurred around 1.85 billion years ago, and is about the same size as the Moon's Albategnius crater. Over 200 million years ago, an impact in Quebec created the circular Manicouagan Lake, a natural annular reservoir measuring 100km across – exactly the same diameter as the lunar walled plain Plato. The feature has a central uplift which was probably formed by the elastic rebound of the Earth's crust immediately after the crater was formed. The youngest of Canada's impact craters is probably the Merewether Crater on the peninsula west of Ungava bay, a 200m-diameter feature which was formed less than 10,000 years ago.

Australian Excavations

The continent of Australia – flat, relatively featureless and tectonically untroubled – is home to a number of astroblemes, all granted a status

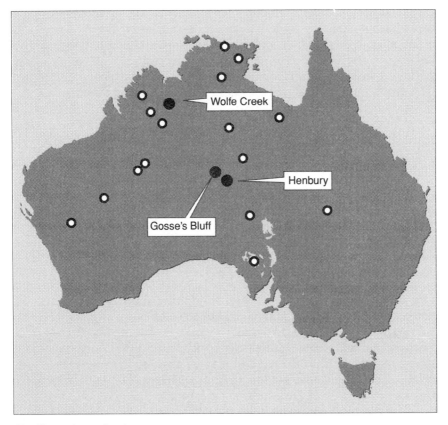

Significant Australian impact craters.

they might never have attained had they been formed on another continent. Pre-dating the Barringer Crater by some 250,000 years, a beauti-
fully proportioned impact crater, discovered by oil prospectors in 1947,
resides at Wolfe Creek in Western Australia. It is somewhat smaller than
its American counterpart, measuring 859m across with a floor 30m deep.
Like Barringer, there are some remaining traces of the original impacting
meteoroid, and the feature is the second-largest impact crater in the
world known to be associated with such meteoritic material. The Wolfe
Creek impactor is likely to have slammed into the desert rocks just as
fast as the Barringer impactor, but it was proportionately smaller.

Just south of Alice Springs, the famous Henbury crater cluster (discovered in 1931) encompasses an oval area of some 1.25sq km and is made
up of 13 craters. The largest of the group measures 220 x 110m, with a
floor up to 15m deep. Because these craters lie in close proximity, the
impacting body must have disintegrated at a low altitude and ploughed
into the landscape from a southwesterly direction. Pieces of meteoritic
material lie strewn about the impact site. The Henbury impact may have

happened within the last 10,000 years; if the craters are so recent, their formation would have made a deep impression upon the region's early inhabitants. It is probably no coincidence that Aboriginal legend has named the site of the Henbury craters 'sun walk fire devil rock', an apparent commemoration of an event which must have been one of the most awesome terrestrial catastrophes ever witnessed by human eyes.

The number of known and suspected terrestrial impact formations is rising, due to the increasing sophistication of geological investigation methods, including surveys from the vantage point of Earth orbit. Only 48 astroblemes were listed in 1972, rising to 91 a decade later. The current tally is around 110, and no doubt many more will be detected in the years to come.

COLLISION: EARTH!

The orbits of over 4,000 asteroids are known with sufficient accuracy for them to have been designated a name and number. New asteroids are being discovered virtually every month; most of them belong to the main zone of asteroids between Mars and Jupiter and have been detected photographically by large professional observatories.

As soon as a new asteroid's orbit has been determined accurately and it has been observed on at least three consecutive or four separate oppositions, the object is numbered and may be given any name (within reason) by the finder. Consequently, there are asteroids named after famous personalities in science and the arts; even totally fictional characters like Mr Spock have gained heavenly immortality.

On 4 October 1989, amateur astronomer Brian Manning of Kidderminster, near Birmingham, secured a photograph of a tiny 16th magnitude asteroid initially designated 1989 TE. After further observations, plus confirmation found on photographs which showed the same object in 1968, the discovery was promoted and given the number 4506 – officially, the first asteroid found from Britain to have had this honour since one discovered at Greenwich Observatory on 5 October 1909. Long before his discovery, Manning's devotion to the field of astrometry (precise positional measurements of astronomical objects, notably comets) was recognized when asteroid 3698 was named in his honour.

In January 1991, Manning photographed another asteroid which climbed the ranks to achieve individual named status. In a citation in the *Minor Planet Circular* for April 1991, Manning's wife Alice was recognized as the real motive force behind the astronomer's impressive achievements:

Minor Planet 4751 Alicemanning. Named in honour of Alice K.
Manning, wife of the discoverer, for her support and encouragement
of the discoverer's hobby over many years. The attention given to

practical details such as warm clothing on cold nights, and the push needed to persuade him to finish the construction and fitting of measuring-machine encoders has contributed greatly to cometary astronomy and success in discovering minor planets.

Astronomers have identified a small percentage of newly discovered asteroids which belong to three groups that can approach perilously close to our planet. Each of these near-Earth groups has its own peculiar type of orbit around the Sun.

Much of the credit for tracking down elusive near-Earth objects must be given to the American astronomer Eleanor Helin, who was instrumental in starting Palomar Observatory's highly successful Planet-Crossing Asteroid Program in January 1973 – at a time when only a handful of asteroids outside the main asteroid belt were known.

Most of the members of the Amor, Apollo and Aten near-Earth groups (each named after a prominent group member) are probably objects which have been gravitationally purloined from the main asteroid belt between Jupiter and Mars. Measurements have shown the near-Earth asteroids to have a good variety of compositions, which means that they are liable to have been derived from a number of different types of parent body.

Amor Asteroids

Members of the Amor group have orbits which take them from the main asteroid belt into the inner solar system between the orbit of Earth and Mars. Asteroid Amor itself was discovered in 1932, but the first Earth-grazing asteroid and member of the Amor group was photographically discovered at the Urania Observatory in Berlin in 1898 and named Eros, after the Greek god of love. Eros has been found to have dimensions of 14 x 27km, with an irregular surface which is brighter on one side than the other. The asteroid can approach to within 25 million km of the Earth, when it becomes bright enough (around the seventh magnitude) to be spotted in binoculars. Consequently, Eros undergoes constant gravitational perturbations caused by the Earth.

In 1931, an international team made a series of precise measurements of Eros' position against the starry background, enabling an accurate determination of the scale of the solar system to be made by means of parallax. A simple demonstration of parallax can be made by switching the view of an outstretched fingertip between your right and left eye and noting its changing position against the background; your fingertip is a very close Eros, and the right and left eyes are two observing stations located on either side of the Earth.

Asteroid 253 Mathilde was discovered at the telescope eyepiece in the late nineteenth century and added to the ever-growing list of objects that circle the Sun in the main asteroid belt between the orbits of Mars and Jupiter. Viewed from the Earth, Mathilde appears as a faint star-like object, even through large telescopes; it sometimes approaches the Earth to within 1AU (150 million km), and is clearly no threat to our planet in such an orbit. On 27 June 1997, NASA's NEAR (Near-Earth Asteroid Rendezvous) spacecraft underwent a close encounter with Mathilde, viewing the 59 x 47km-diameter asteroid at close range and obtaining more than 500 stunning images.

NEAR's imager revealed a highly cratered asteroid. Cosmic collision, it appears, has played a major role in shaping Mathilde, the asteroid having suffered extensively from big meteoroidal impacts. Its dark surface displays five craters bigger than 20km across, the largest of which is more than three-quarters of the asteroid's diameter and estimated to be at least 10km deep. Proportionally, this crater is the largest impact feature on any body in the solar system, bigger even than crater Stickney on the Martian moon Phobos; scientists joked that at first glance Mathilde boasts more crater than asteroid! It remains a real mystery how Mathilde has managed to survive intact after so many mighty impacts, because such massive collisions in relation to Mathilde's overall size should have smashed the asteroid to bits.

Mathilde is a very dark, carbon-rich, C-type asteroid with an albedo (reflectivity) of only 3 per cent, making it more than twice as dark as the tarmac-grey lunar plains. The asteroid's pitch-black colouration appears uniformly throughout the object; there is no variation in colour, even at depths excavated by the big impacts. Mathilde does not appear ever to have been part of a larger body, and is likely to be composed of material which coalesced at the birth of the solar system. The asteroid is far less dense than had originally been thought, probably a compact conglomeration of smaller bodies of identical primordial composition (if Mathilde was just a chunk of a bigger asteroid which broke up, its density would be expected to be far higher than it actually is). The stuff of Mathilde may be representative of the sorts of debris that originally accreted to form the planets, since it has not been altered in any significant way over this vast period of time by planet-building processes. Closer analysis of the asteroid by future probes would undoubtedly throw a great deal of light upon the physical and chemical condition of the early solar nebula some 4.5 billion years ago, providing science with its long-sought interplanetary Rosetta Stone.

After this most successful encounter – the first of NASA's 'cheaper, faster, better' Discovery Class missions and by far the least expensive interplanetary fly-by in NASA's history – NEAR's thrusters fired and the craft was sent back towards the Earth for a gravity-assist which will ultimately send the little probe to the Amor asteroid Eros in early 1999, completing an interplanetary trek of 2 billion km. NEAR is expected to remain in orbit around Eros for more than a year, from where it will make a series of detailed studies of the asteroid.

Apollo Asteroids – and a Near-Miss

More than 50 Apollo asteroids have been identified. These objects have orbits which take them from the main asteroid belt to plunge further into the inner solar system than the Amor members, crossing the orbit of the Earth itself and presenting a far greater risk of colliding with our planet at some point in the future. Many of the asteroidal scars visible on the Earth, Moon and Mars have been produced by Amor and Apollo asteroids whose route has been blocked by the planet in question. Most meteorites found on Earth are probably small chunks of material knocked off Apollo group asteroids; many chondrite meteorites, for example, display a strong compositional resemblance to the asteroid Apollo.

On 25 October 1937, the tiny 1km-diameter Hermes – probably the best known of all the Apollo asteroids – was discovered as an elongated streak on a photographic plate. Five days later the diminutive asteroid had sped by the Earth at a distance of just 780,000km, around twice the Moon's distance, and was visible through binoculars as a magnitude 8 star whose motion against the stars – a staggering 5 degrees per hour – could actually be observed in real-time. It took Hermes just nine days to cross the entire sky in its rapid interplanetary fly-by. This event sparked the first modern concern for the Earth's future in the face of an extra-terrestrial threat. Newspapers ran headlines which announced that the end of the world had only just been narrowly avoided; Hermes had crossed the Earth's orbit within just five hours of the Earth having been in the same position. This distance becomes incredibly tiny if it is converted to the speed of light – we came just three light seconds away from a global disaster.

In Britain, the *Sunday Graphic* reported Hermes' encounter more than two months after the asteroid had vanished into interplanetary space, but the newspaper still ran the startling headline story which stated:

WORLD DISASTER MISSED – By 5 Hours
Scientists Watch a Planet Hurtling Earthward

The Earth has just escaped disaster and possibly destruction in a collision with another planet by five and a half hours. Astronomers in England and South Africa are the witnesses.

'It was the nearest thing to a collision between the earth and another body that has ever happened,' said Dr H. E. Wood, South African astronomer, in Cape Town yesterday. He revealed that he and 30 other Cape astronomers had watched the phenomenon.

On October 25 a minor planet was first seen to be rushing through space, making for the Earth in a straight line. As the days passed it was obvious that if the planet missed us it would only be by an astronomical fraction. Finally it shot past the earth, 400,000 miles and five and a half hours away – a small astronomical distance.

Narrowest Escape

English astronomers who also observed the phenomenon spoke guardedly, but they agreed yesterday that the world had the narrowest escape ever recorded. Mr William M. H. Graves, one of the two secretaries of the Royal Astronomical Society, told the *Sunday Graphic*: 'According to our observations, it was about 600,000 miles away at its nearest position. Shortly before the war a big meteor [actually, a reference to the Tunguska cometary impact] fell in a deserted region of Siberia and damaged several hundred square miles.'

Exactly a year after Hermes' near-collision, Orson Welles panicked a substantial proportion of the US population in his CBS broadcast of *The War of the Worlds* – perhaps the actor's decision to produce the play had been influenced in no small manner by the way in which people seemed to have been shaken by the news of Hermes' threat to human civilization. Welles' broadcast was, thankfully, a piece of dramatic fiction. But Hermes remains out there somewhere, just as potentially dangerous as it was in 1937 – one of many cosmic threats against which the Earth is largely defenceless.

Aten Asteroids

The third group of near-Earth asteroids have orbits which lie largely within the Earth's but intersect us on the far leg of their orbit. The first such asteroid was discovered by Helin in 1976 and named Aten; since then, only a few members of the Aten group have been identified, but the search goes on. In all, it has been estimated that around 2,000 near-Earth asteroids larger than 1km in diameter lie waiting to be discovered.

Recent Earth-Grazers

On 29 November 1996, the asteroid Toutatis (designated asteroid 4179) passed within 3.5 million km of the Earth – about ten times the distance of the Moon, which is uncomfortably close by interplanetary standards. Radar images of the near-Earth asteroid were obtained in December 1992, during its previous close encounter with a signal beamed by the 70m dish at Goldstone in California and received by the new 34m antenna. The returning radio signals were converted into remarkably detailed images which showed a double asteroid measuring around 8 x 5km whose components were in contact with each other, both being pockmarked with impact craters. The largest of these was of comparable size to the Earth's Barringer Crater. Toutatis, first observed in 1934 by Eugene Delporte, has an orbital plane inclined just 0.5 degree to that of the Earth, and will continue approaching us long into the future. It may ultimately be used as a convenient staging post for manned space activities in the twenty-first century and beyond.

On 31 March 1989, a tiny asteroid called 1989 FC passed within

690,000km of the Earth – smashing Hermes' record set more than half a century before – and reportedly came less than seven hours away from collision. Since it had approached from the direction of the Sun, the 300m-diameter asteroid was not spotted until it had passed the Earth and was identified on photographs taken by Henry Holt and Norman Thomas using the 450mm Schmidt telescope on Mount Palomar. It was then a star-like point at a very faint magnitude 16.5 – far too faint to be seen without the aid of a large telescope. Astronomers calculated that 1989 FC has an orbital plane inclined some 5 degrees to the Earth's (making the likelihood of collision more remote) and that the object orbits the Sun in an eccentric orbit, approaching the Sun closer than Venus, in just over a year. The asteroid approaches close to the Earth every 20 years or so, and its next close pass will take place in 2012. If 1989 FC had impacted on the Earth it would have exploded with a force equivalent to 20 billion tonnes of TNT, carving out a crater at least 10km across and 1km deep.

On 17 January 1991, the near-Earth distance record was broken yet again by the asteroid 1991 BA, a 100m-wide lump of iron discovered at the University of Arizona, which passed within 170,000km of the Earth – less than half the distance of the Moon. The year ended with another close asteroidal encounter when, on 5 December 1991, the asteroid (more appropriately, a large meteoroid) 1991 VG, discovered a month earlier at Kitt Peak Observatory near Tucson, Arizona, passed about 460,000km from the Earth. The object was found to move in an orbit similar to the Earth's and had a brightness curve consistent with it being an asteroid rather than a large unidentified rocket stage, which some had claimed it to be. Attempts by a team at the NASA tracking antenna at Goldstone to bounce radio signals off the object in order to determine whether or not it was metallic were not successful, and 1991 VG remains mysterious. The object is probably a meteoroid measuring under 10m from end to end, which orbits in a path that will bring it into our vicinity again in the year 2008.

Face-to-face with a Mile-wide Asteroid

In less than 30 years, inhabitants of the planet Earth will experience the closest encounter in recorded history. A massive lump of rock, named 1997 XF11, 1.6km in diameter, was discovered in December 1997 by University of Arizona astronomer, Jim Scotti, at Kitt Peak Observatory. Scotti's astounding find was made as part of his duties in conducting the Spacewatch programme, a testament to the success of vigilant space monitoring.

After measuring XF11's path against the stars, an accurate orbit was quickly calculated. Astronomers were astonished to find that XF11 will make an exceptionally close approach to the Earth in October 2028. The exact time and distance of this encounter is still not known precisely because more observations of the asteroid's motion are required over the coming years to refine calculations. Although it is currently a possibility, the asteroid is thought unlikely to be on a collision course for the Earth – this time around! One reasonable calculation has determined that XF11 will speed by the Earth at a distance of a mere 40,000km on the evening of 26 October 2028. Ten times closer than the Moon, the asteroid will be clearly visible from Europe as a brilliant star, zipping through the night sky from northwest to southeast during the course of a few hours, if these calculations are correct. The shape of the asteroid will be clearly visible through binoculars and its cratered surface will be discernible through guided telescopes. Before the main event, an ideal opportunity to spot XF11 as a small star will be on Halloween in 2002, when the asteroid approaches the Earth at a distance of some 9.6 million km.

Selected Near-Earth Approaches by Asteroids 1998 to 2005

Asteroid	Date	Distance (AU)	Magnitude
3361 Orpheus	12 Feb 1998	0.1668	19.0
6037 1988 EG	28 Feb 1998	0.0318	18.7
1994 AH2	17 Jun 1998	0.1930	16.5
1987 OA	18 Aug 1998	0.1019	18.5
1991 RB	18 Sep 1998	0.0401	19.0
1865 Cerberus	24 Nov 1998	0.1634	16.8
1989 UR	28 Nov 1998	0.0800	18.0
1992 SK	26 Mar 1999	0.0560	17.5
1991 JX	2 Jun 1999	0.0501	18.5
4486 Mithra	14 Aug 2000	0.0466	15.4
4179 Toutatis	31 Oct 2000	0.0739	15.4
1991 VK	16 Jan 2002	0.0718	17.0
4660 Nereus	22 Jan 2002	0.0290	18.3
5604 1992 FE	22 Jun 2002	0.0768	17.0
1990 SM	17 Feb 2003	0.0747	16.5
1991 JX	20 May 2003	0.0923	18.5
1994 PM	16 Aug 2003	0.0246	17.5
1990 OS	10 Nov 2003	0.0304	20.0
1995 CR	29 Dec 2003	0.0637	21.5
6239 1989 QF	2 Feb 2004	0.0564	18.1
4179 Toutatis	29 Sep 2004	0.0104	15.4
(closest approach: 1,555,000km)			
1988 XB	21 Nov 2004	0.0729	17.5
1992 BF	3 Mar 2005	0.0630	19.0
1993 VW	24 Apr 2005	0.0862	16.5
1992 UY4	8 Aug 2005	0.0402	17.5
1991 RB	13 Sep 2005	0.0785	19.0
1862 Apollo	6 Nov 2005	0.0752	16.3

(AU – Astronomical Unit – equal to the Earth's average distance from the Sun, some 149,598,000km)

Cometary Close Shaves

When the long-lost Comet Swift-Tuttle was rediscovered in 1992, comet expert Brian Marsden claimed his calculations showed that the object would make a very close pass of the Earth at its next visit in the year 2126. Marsden went so far as to predict that there was one chance in 10,000 that Swift-Tuttle's nucleus would collide with the Earth on that occasion. The world's media pounced on this suggestion, leading to numerous end-of-the-world stories, like those inspired by asteroid Hermes' approach more than half a century earlier. Whatever happens in 2126, the comet will make a very spectacular sight because of its extreme proximity, and there is the added possibility that it will be accompanied by a magnificent display of Perseid meteors, a stream of debris which is known to have been deposited in space by the comet.

Two recent comets have come very close to the Earth. During a few days in May 1983, Comet Iras-Araki-Alcock sped through the stars of Draco, almost overhead across Ursa Minor, and then proceeded through Ursa Major towards the southwestern horizon. Iras-Araki-Alcock was easily visible to the naked eye as a hazy, circular patch of light some 2 degrees across (four times the apparent width of the Moon) with a bright nucleus offset to one side.

Iras-Araki-Alcock's nucleus, measured with infrared observations to be around 5km in diameter, passed just 4.6 million km from the Earth on 11 to 12 May. At this time, the comet's movement against the background stars could be detected without difficulty; like hundreds of amateur astronomers, I took sheer delight in spending hours plotting the comet's path on a star map. It is not often that a celestial spectacle occurs in such a way that a feeling of depth is conveyed to the viewer, but on this occasion the solar system three-dimensionalized dramatically before our very eyes.

Comet Hyakutake of 1996 was the brightest comet to appear in British skies since Comet West of 1976 (until it was surpassed by Comet Hale-Bopp which turned up a year later, in 1997). Hyakutake was the nearest bright comet to approach this close to the Earth in 400 years, and it made a splendid showing during March and April 1996, when it raced across the northern constellations of Boötes, Ursa Minor and Perseus. The visitor passed the Earth at a distance of just 15 million km and then brightened as it approached the Sun. Some predictions of Hyakutake's brightness were a little too optimistic; one eager astronomer reckoned it might become as bright as Venus. In the end, though, the comet did not disappoint, becoming an easy naked-eye object even from light-polluted urban sites. Hyakutake developed a considerable tail over 20 degrees long and had a curious greenish glow which was obvious through binoculars.

IMPACT DEVASTATION

As they are so large and have such a high velocity, comets and asteroids are capable of wreaking immense planet-wide destruction many orders of magnitude worse than a total global nuclear war. An iron-rich meteoroid 100m in diameter, impacting on the Earth at a velocity of some 19km per second, would carve out a crater at least 2km across and the resulting fireball and shock waves would destroy all life within 200sq km around it. Such impacts may occur with the alarming frequency of perhaps once every 22,000 years.

A 500m-diameter meteoroid will produce a crater more than 10km wide and wipe out anything within 1,000sq km around the impact site. Big meteoroids like this might hit our planet every half-million years or more, producing environmental catastrophes which permanently affect life the world over. An asteroid ten times larger will blast out a crater in the Earth's crust exceeding 100km in diameter, rendering total desolation over half a million sqare kilometres.

Finally, the unimaginable catastrophe of the impact of a 30km-diameter asteroid would probably mark an abrupt end to human civilization. The object would gouge out a crater (not a deep depression but rather a circular impact zone of crustal fissures and faults) more than 200km across and obliterate a surrounding area of 90 million sq km of terrain. Doomsday asteroids like this have been predicted to collide with the Earth every 4.5 million years, and have been responsible for mass extinctions of animal life.

Preventing Doomsday

Ever since Charles Messier decided to devote his time to spotting new comets (he discovered 14 of them) and Giuseppe Piazzi discovered asteroid Ceres (on the very first day of the nineteenth century), comet and asteroid hunting has been regarded as something of a specialized scientific sport, requiring skill, patience and a certain amount of luck. Piazzi was a member of the self-appointed 'Celestial Police' – a group of astronomers determined to populate an area of space, once considered empty, with their new minor planetary discoveries.

Mankind is waking up to the fact that the tiny, benign-looking asteroids seen through the telescope eyepiece or as dim smudges on photographic plates are actually huge mountains of solid rock and metal which on occasion pass dangerously close to the Earth. Today, with an increasing number of Earth-grazers being discovered by luck or by design, the asteroid-hunting game remains the same but the stakes are far higher.

The International Near-Earth Asteroid Search was established in 1984 by Eleanor Helin, and consists of a network of wide-field Schmidt telescopes located in Europe, Japan and the US. The entire network is alerted immediately when one camera detects a fast-moving object; its position is relayed and accurate co-ordinated observations help to determine the exact orbit of the object. Scientists hope that this strategy will prevent another body like Hermes from taking everyone by surprise and vanishing before its path can be fixed with any degree of certainty. By 1989, the wide-field Spacewatch Camera at Kitt Peak Observatory alone had pinpointed 25 comets and found a staggering 250 new asteroids.

The twenty-first century is likely to see the formation of a celestial police force with the power to make an arrest rather than simply take note of a rapidly receding cosmic number-plate. It is not difficult to imagine a sophisticated detection system capable of spotting all sizeable asteroids and comets long before they approach the Earth. An integrated network of ground-based and orbiting wide-field telescopes combined with infrared satellites would be a costly but effective means of achieving this end.

An object approaching Earth from the general direction of the Sun could be detected using such equipment, situated at the relatively stable Lagrangian orbital points 60 degrees preceding and following the Earth as measured from the Sun. Any large object which threatens our planet, or for that matter the Moon or Mars, both of which may host permanently manned bases, will either be destroyed at a safe distance or diverted from its course into a more benign orbit.

In the disaster movie *Meteor* (1979), Hollywood portrayed the menace of a large asteroid (definitely not the 'meteor' of the title!) heading directly towards the Earth. Even at the height of the cold war, Russia and America decide to co-operate with each other by combining their secret orbiting nuclear arsenals and directing the missiles toward the incoming object. The asteroid is fragmented by the detonations but several large chunks happen to survive and impact around the world, causing a series of terrible disasters.

Such an approach to the problem is drastic and probably unnecessary; it would be more effective to deal with fragile cometary nuclei rather than stronger asteroids in this way. Instead, it might be wiser to explode nuclear weapons several asteroid-diameters from the body so that the shock waves produce a small but vital change in the asteroid's velocity, nudging its path away from imminent impact with the Earth. Unless it is slowed down considerably, a massive, fast-moving asteroid is unlikely then to enter into orbit around the Earth as another satellite; instead, it would make a spectacular slingshot around the Earth and enter a new orbit around the Sun, whose configuration would depend on the circumstances produced by orbital mechanics.

Solenoid Asteroids – Magnetic Fantasy?

In 1809, the Danish physicist and chemist, Hans Christian Olmsted, discovered that a compass needle is deflected when it is in proximity to a wire carrying an electric current. The electric current produces a magnetic field around the wire, a discovery which marked the beginning of scientific electromagnetic studies. If an insulated wire is coiled around an iron core and an electric current is passed through the wire, a strong magnetic field is produced as the particles of iron constituting the core align themselves in the direction of the magnetic field. This is the basis of an electromagnet known as a solenoid, a device at its most potent when it contains a soft iron core which responds well to the magnetic field but rapidly loses its magnetic properties once the electric current is switched off. If the iron core is 'U'-shaped with both arms wrapped in wire, then its power is strengthened as the poles are brought closer together.

Implausible though it may seem, an iron asteroid coiled from end to end in high-voltage insulated electric cable would act precisely as a giant electromagnet, whose power could be provided by a nuclear reactor. If a central section of the asteroid were removed, so that it resembled a doughnut with a bite taken out of one side, and both lobes were coiled with cable, then the asteroid would act as a more powerful horseshoe magnet. Such an object might be a useful tool in altering the paths of other iron-rich asteroids threatening our planet. After powering-up the nuclear engines, the trick would be to direct the solenoid asteroid to an encounter with the aggressor as far as possible from the Earth.

The huge asteroidal electromagnet would be powered-up as it approached close to the iron asteroid. A head-on encounter would be of too short a duration to be of much consequence. The most effective approach of the solenoid would be to match the asteroid's path and velocity closely in a fly-by and then switch on the power to effect a magnetic docking between solenoid and aggressor. The nuclear engines would then be fired at full throttle towards the bodies' direction of travel; even a relatively small change in the objects' path would produce a significant alteration further down the line. Alternatively, the nuclear reactor could be encouraged to explode to ensure an effective alteration in its path. Likely sites for these batteries of orbiting solenoid asteroids would be at the gravitationally stable Lagrangian points around the orbits of the Earth, Moon and Mars, from where they could be launched in the event of an emergency.

THE GREAT JOVIAN COMET CRASH

In the summer of 1994, one of the most spectacular astronomical events to occur in our solar system for many hundreds of years astonished astronomers the world over. From 16 to 22 July, more than 21 pieces of the fragmented nucleus of Comet Shoemaker-Levy 9 – ranging in size from lumps 600m across to more than 4km in diameter – slammed one by one into the atmosphere of giant planet Jupiter. More telescopes were aimed at Jupiter during that week than had ever been pointed at any other area of the sky in nearly four centuries of telescopic history.

Comet Shoemaker-Levy 9 (SL9) was discovered in March 1993 by American astronomers Eugene and Carolyn Shoemaker and David Levy; the object represented the team's ninth joint comet discovery. After careful observations, it was calculated that the comet had once been in a distant near-circular orbit around the Sun, in the ecliptic plane between the orbits of Jupiter and Saturn. Around the year 1929 (the very year of Carolyn Shoemaker's birth), SL9 was captured by Jupiter's gravity and assumed a near-polar orbit around the giant planet, making one revolution around it in 2 to 2.5 years. The comet made several close approaches to Jupiter in the ensuing years, notably in 1940–2 and in 1970. In July 1992, eight months before its discovery, the comet approached the cloud tops of Jupiter to within a distance of only 43,000km, ten times closer than the innermost Galilean satellite, Io. Consequently, the immense gravitational forces SL9 endured during this very close pass were enough to break the single icy nucleus into many smaller pieces.

When SL9 was imaged closely, it assumed a peculiar string-of-pearls appearance in the midst of the faint haze of its dusty coma, a mass of unresolvable debris arising from the fragmentation. No gaseous emissions were detected from the comet, an observation that did not surprise scientists, since the solar energy received by SL9 at that distance was too weak to produce noticeable activity.

Many months before the fragments of SL9's nucleus were predicted to hit Jupiter's southern hemisphere, astronomers began to speculate about the likely observability of the effects of the impacts. The entry of each cometary nucleus into the Jovian atmosphere would be preceded by a magnificent coma meteor storm, as billions of dust particles burned up high above Jupiter's clouds. From the Earth, each meteor storm would be observed as a flaring event on Jupiter's limb that lasted a few seconds. From a hypothetical view within Jupiter's atmosphere, the scene would have been so awesome as to put any terrestrial meteor storm in the shade! It was agreed that the impacts themselves would not be directly visible from the Earth, since the cometary fragments would arrive on Jupiter's far-side, just past the visible edge of the planet. However, the rapid rotation of Jupiter (the giant planet's day is less than ten hours long) would quickly bring the impact sites into view. If there were any scars visible in Jupiter's clouds, then they would rotate on to the Jovian near-side less than half an hour after their formation.

Some astronomers predicted that the brilliant fireballs produced during the impacts on Jupiter's far-side would be so dazzling that they would noticeably illuminate the faces of the Galilean satellites Io, Europa, Ganymede and Callisto. Some hoped that the fireballs themselves would pop up past the edge of Jupiter and appear as bright blobs of light rising slowly above the planet's limb. Other high expectations included permanent atmospheric changes brought about by the deep disturbances, including the possibility that Jupiter might develop a new set of cloud belts or even a new Great Red Spot (the old Great Red Spot is a huge, revolving storm that has been raging in Jupiter's southern hemisphere since the mid-nineteenth century).

Most astronomers were cautious. Like many others, I guessed that through average-sized telescopes few visible effects would be caused by the impacts. Weeks before the event I had convinced myself that a few 1km-sized chunks of ice could not possibly produce any noticeable effect on the giant 143,200km-diameter Jupiter. When the British astronomer Patrick Moore reckoned that the SL9 impacts would have as little effect as throwing a handful of baked beans at a charging rhinoceros, many of us imagined that even this analogy would prove to be a wild exaggeration in the comet's favour!

Those who saw the effects of SL9's smash into the clouds of Jupiter witnessed a spectacle the like of which is unlikely to be seen again for a very long time. It was an awe-inspiring sight which amateur astronomers will never forget. On the evening of 20 July I set up my 100mm refractor, waiting for the skies to darken and for Jupiter to come into view in the southwest. Once the sky had darkened sufficiently I located Jupiter, trained my telescope towards it and focused the image. The most incredible sight presented itself in the field of view. Mighty Jupiter – the biggest planet in the solar system – appeared visibly scarred by fragments of impacting cometary nuclei. The impact scars took the form of intense black spots surrounded by shock-wave patterns. The most prominent scar, caused by the impact of 4km-diameter 'fragment G', took the form of a black blob bordered on its preceding edge and to the south by a semicircle of dark material. Over the next few days the scene on Jupiter constantly changed before the eyes of astronomers, as further impacts created new atmospheric disturbances and old ones developed into complicated shapes in the southern temperate region.

Over the months following the SL9 impacts, astronomers eagerly followed Jupiter into the evening twilight as the planet headed for conjunction with the Sun in November. When Jupiter had rounded the Sun and emerged into December's morning skies, the effects of the impacts which had happened six months before were still noticeable through small telescopes. Material that had been brought to the cloud surface from deep within the Jovian atmosphere had spread around the latitude of the impacts, and some traces of the old impact sites were still detectable within this new dusky 'south south temperate' belt. Jupiter's atmosphere has now healed and no longer shows any signs of having been assaulted by a volley of mountain-sized cometary nuclei.

Now that astronomers know exactly how Jupiter is affected by such cosmic impacts, research into old astronomical archives has brought to light several examples of unusual Jovian atmospheric phenomena observed throughout history. In December 1690, the Italian-born French astronomer Giovanni Cassini (who discovered four Saturnian satellites and the apparent gap of the 'Cassini Division' in Saturn's rings) observed a single black spot in Jupiter's equatorial region which, over the next 18 days, developed in complexity and spread over a portion of that latitude.

It has been predicted that Jupiter experiences the impact of a 1.5km-diameter object once every 500 to 1,000 years. Gene Shoemaker calculates the rates of impact to be higher, with one such cometary collision happening every century. However, events like the impact of a pre-fragmented cometary nucleus such as SL9 are much rarer, occurring maybe once every 2,000 years or so. We know that asteroids and comets have strayed too close to Jupiter in the past, because they have impacted on Jupiter's moons and left permanent scars. For example, a number of crater chains on the surface of icy Callisto, Jupiter's outermost large satellite, show that objects have been broken up by Jupiter's immense gravity and have slammed into Callisto's surface in rapid succession. The diameters of these relatively fresh craters indicate that the impactors were each around 400m to 800m in diameter and numbered 6 to 25 fragments per impact event.

The SL9 impact prompted the US Congress to direct NASA to look into the setting-up of long-term programmes to identify comets and asteroids that may pose a threat to the Earth. At present it is only feasible to identify asteroidal bodies that are larger than 1km across and likely to cross the Earth's orbit within the next decade. Warmer and brighter cometary bodies with smaller nuclei may be detected further afield in the visual and infrared frequencies. Congress authorized NASA's collaboration with the Department of Defense, a pairing which worked well with the Clementine lunar probe of 1994. It also places a less severe strain on both agencies' budgets.

The DoD has a telescopic deep-space surveillance programme devoted to tracking satellites and man-made debris in Earth orbit. With suitable modifications to the computer software, this sophisticated set-up could be employed to search automatically for asteroids or comets (which usually appear to travel much more slowly against the starry background), in addition to operating its current defense programme. DoD satellites looking down on the Earth have the capability of spotting bright meteor flashes and the re-entry of space hardware and space junk – these sensitive instruments have been returning valuable data on the frequency of cosmic impacts. An eight-member committee headed by Eugene Shoemaker was appointed by NASA to determine a workable plan.

IMPACT SCARS OF THE RED PLANET

Our distant ancestors beheld the blood-red point of light we now know as planet Mars and, naturally enough, associated it with their gods of war. In July 1965, the US spaceprobe Mariner 4 showed us that the ruddy face of warrior Mars is extensively battle scarred with the marks of billions of years of cosmic impact. When the two US Viking probes sampled the Martian rocks in 1976, they demonstrated that Mars is the reddest permanent object in the heavens because frozen water locked in its rocks has caused the iron minerals to oxidize – put simply, Mars is red because its surface is slowly rusting away.

Mars has a diameter of 6,790km, a little larger than half the Earth's diameter. Through a telescope, the Martian disc always appears more than 85 per cent illuminated because we are viewing it from a position in the inner solar system. However, Mars is not a big planet and it never becomes spectacularly large in the eyepiece. When Mars appears opposite the Sun in the sky, riding high above the southern horizon at midnight, the planet is said to be at 'opposition', its closest point to us in that particular orbit.

Mars approaches Earth during these opposition periods every 780 days, but because its orbit is rather elliptical, its opposition distance from us can actually vary between 100 million and 56 million km. Every 15 years there are two close perihelic oppositions (when Mars is nearest to the Sun) and five more distant oppositions. In the telescope eyepiece, Mars can attain a maximum apparent diameter of nearly 25 seconds of arc; useful observations with medium-sized telescopes are viable only when the planet exceeds 10 seconds of arc.

In November 1659, the Dutch astronomer Christian Huyghens made the first drawing of Mars depicting recognizable surface features, including a prominent V-shaped, dusky marking known as the Syrtis Major. In 1666, the French astronomer Giovanni Cassini discovered white patches at the planet's poles, and in 1719 Giovanni Maraldi, Cassini's one-time assistant, correctly inferred that these patches were extensive tracts of polar ice.

Astronomers soon concluded that most of the Martian markings are more or less permanent surface features, although they appear to vary somewhat in shape and intensity from season to season. Until the surface of Mars was scrutinized by spaceprobes, astronomers did not agree among themselves about what the markings might represent. Syrtis Major can still be seen today, but in two centuries of speculation this marking has been variously diagnosed as a deep geological valley; a raised mountain plateau; a wide, dark Martian ocean with an area larger than the Mediterranean Sea; and a tropical forest of Amazonian proportions. None of these has proved correct.

That Syrtis Major and many other of the dusky tracts appear to undergo seasonal fluctuations in intensity was an observation central to the arguments of those who claimed that the dark areas were zones rich in vegetation. Sadly, the truth is more mundane. Mars' desert markings are actually areas of darker crust which are overlain by sand dunes. As the Martian seasons progress, winds shift the sands to expose and obscure alternately the underlying surface, producing a variation in overall intensity. Although noticeable from Earth, the cause of this can only be resolved at close quarters.

The astronomer Giovanni Schiaparelli is responsible for having commenced Mars' rather exotic nomenclature; he based it upon ancient Mediterranean and Middle-Eastern seas and lands, Biblical names and a little classical mythology thrown in for good measure. Before Schiaparelli's names were accepted, some astronomers took it upon themselves to name the red planet's features after their own fancies – just as the Moon-mappers of old had done centuries before.

The Martians and their Canals

An almost continuous tract of prominent dark areas encircles Mars and takes up most of the planet's southern hemisphere. Schiaparelli's map of Mars, made after the 1881–2 apparition, appears as an intricate patchwork of dusky promontories and spots connected to each other by a network of long lines, many of which are double. Schiaparelli named these lines *canali* (Latin for 'channels'), and imagined them to be planet-wide features with a natural origin, perhaps faulting. However, the imagination of other astronomers had been fired, and speculation that the *canali* were canals constructed by intelligent Martian life-forms became rampant at the end of the nineteenth century. Soon, just about every telescopic observer was seeing the canals of Mars, regardless of the size of telescope employed, the seeing conditions or the apparent size of the Martian disc! (For further details, see page 17.)

The greatest proponent of the canals of Mars (and the aliens imagined to have constructed them) was the wealthy American businessman Percival Lowell. He wrote several bestselling books on the subject and constructed an observatory dedicated to the study of the red planet atop a hill at Flagstaff, Arizona. The observatory was equipped with a superb 600mm refractor, and with it Lowell spent countless hours viewing Mars and drawing the features he thought he could see. Lowell's maps of Mars are criss-crossed with intricate networks of fine lines. These, Lowell asserted, were areas of vegetation bordering canals which carried fresh water from the poles down to the arid equatorial regions, and were the magnificent handiwork of a peaceful, well-ordered race of Martians.

Some astronomers challenged the reality of the Martian canals. Eugene Antoniadi, one of the greatest visual observers who ever looked through a telescope, accepted that features resembling canals could be seen on the Martian disc. However, with a good telescope and an acute eye, under excellent seeing conditions (and Antoniadi had all of these advantages), the linear features break down into a mass of finer detail. Chance alignments of dusky patches and areas of differing contrast combine to create the illusion of lines.

Now that spaceprobes have mapped Mars, we know that Antoniadi was right. The Martian canals never existed, and neither did the race of intelligent Martian engineers. The fact that so many people saw the canals in the late nineteenth and early twentieth centuries attests to the power of suggestion, made even more potent by official endorsement. It is a phenomenon worthy of the attention of social scientists and astronomers alike.

Features of Mars

Mars displays a distinct north–south difference through the telescope, with most of the dusky areas occupying the southern hemisphere. Many of these regions, including Syrtis Major, appear to project northwards. Solis Lacus (Lake of the Sun) is an interesting dark patch situated in a brighter, circular region and bordered by a dusky arc – a striking appearance which has earned it the name of the 'Eye of Mars'. Mare Acidalium is the most prominent of the northern hemisphere's dusky tracts. Surprisingly, most of the dusky Martian markings visible through the telescope do not really correspond to areas of pronounced surface relief.

Bordering the south of these darker areas, the bright circular regions of Hellas and Argyre are the remnants of huge ancient impact craters which have been filled with wind-deposited sands. Long ago, when Mars was a wet world, these features are thought to have been full of water. Hellas and Argyre are easily visible through small telescopes during favourable oppositions of Mars, and they can vary in apparent brightness during an apparition. The smaller features of Edom (north of Sinus Sabaeus on the equator) and Oenotria (south of Syrtis Major) are also outlines of large Martian impact craters, now named Schiaparelli and Huyghens respectively.

All these features are visible because of their brightness and contrast with surrounding areas. However, any attempts to see the shadow effects produced by topographic forms must concentrate upon the Martian terminator, the division between the illuminated and unilluminated hemispheres of a planet. This requires an exceptionally keen eye, a good telescope and perfect seeing conditions. Using the superlative 900mm refractor of the Lick Observatory at Mount Hamilton, California, the eagle-eyed American astronomer Edward Barnard observed a great amount of Martian topography during the favourable apparition of 1892–3. Barnard even managed to observe Martian craters, the Tharsis volcanoes and the impressive Mariner Valley (Vallis Marineris), although the true nature of these features only became apparent after spaceprobes had reconnoitred the planet.

Spacecraft have revealed distinct geological and topographical differences between Mars' northern and southern hemispheres. The southern hemisphere is covered with many lunar-type craters, and it averages 3km higher than the northern hemisphere. Extensive volcanic activity has modified the northern hemisphere, and smooth sheets of lava have spread over the older crust, obliterating more ancient features. Many sizeable volcanic shields have grown to tower over their surroundings, including the massive volcanoes of the Tharsis region.

Meteoroidal impact has produced most Martian craters. There are exceptions, as on our own Moon, where volcanic vents assume a crater-like appearance. The most impressive of this kind of volcanic crater (actually several which overlap each other) is situated at the top of Olympus Mons, Mar's biggest volcano with a base 500km across and a height of 24km. The summit craters look clean-cut because they are so young and relatively unweathered.

Mars' older craters have been modified by millions of years of weathering. Their steep interior and exterior slopes are often incised by deep gullies, similar to those seen in the Earth's Barringer Crater, and may have been cut by running water. The crater Arandas in Acidalia Planitia looks remarkably like the lunar crater Alpetragius, with its proportionately large central mountain. What makes Arandas and many other Martian craters interesting is the presence of peculiar 'mud-splash' patterns in the surrounding terrain, which suggest a meteoroid impact on wet or frozen ground. Another unique form of Martian landscape is the streamlining effect seen around craters and other areas of pronounced relief, indicating substantial water erosion in Mars' history.

Incredibly detailed spaceprobe images have provided geologists and planetary scientists with an embarrassment of scientific riches, although sadly there have been no sample return missions to date. The images returned by Mariner and Viking are still being analysed today, and a new fleet of Martian spaceprobes is poised to secure hundreds of times more data than has so far been accumulated. Mars has had a fascinating history, and water once played a major part in it. Although Mars may once have had a vast ocean occupying its northern hemisphere, the red planet is now cold and dry, with the remaining water locked up in its polar caps and beneath the surface in a layer of permafrost.

Mysterious Martian Meteorites

Until recently, theorists found it difficult to envision impact dynamics powerful enough to launch material off bodies with a gravity as substantial as Mars, but it was always a theory favoured over volcanic ejection. Viable computer-generated models of impact-ejection have now been demonstrated. Impacts are capable of blasting material from the Moon, Mars and all of the solar system's rocky bodies, including the one with the highest escape velocity – Earth. Meteoroids derived from all the solid worlds populate space, and the rarest of these found on Earth have come from the Moon and Mars.

In August 1996, NASA made a surprising announcement, which provides us with a neat link between Mars, meteorites and the Earth, with the startling suggestion that primitive life may once have existed on Mars.

A tiny piece of Mars was found in Antarctica's Allan Hills near McMurdo Station in 1984. Labelled ALH84001, the meteorite was not immediately recognized for what it was. Instead, it was stored away in a drawer under an ordinary meteoritic classification. When the specimen was later re-examined and its composition scrutinized closely, it was found to resemble strongly the rocks of Mars, whose chemical composition had been measured by the Viking landers in 1976. Only a dozen meteorites of similar composition have ever been discovered.

A little scientific detective work enabled the history of ALH84001 to be unravelled. It was found once to have been immersed in water on Mars, and organic molecules (carbon-based compounds) had been deposited in small cracks in the rock. Some 16 million years ago a meteoroidal impact on Mars released enough energy to hurl the rock into space, where it remained until it descended through the Earth's atmosphere to land in the icy Antarctic wastes some 13,000 years ago.

On close examination with a powerful electron microscope, startling evidence was found for thread-like Martian microfossils similar in appearance to known terrestrial bacteria, although much smaller. When this news broke in the world's media in August 1996 it caused tremendous excitement, since it was an official acknowledgement of the likelihood that life may have developed independently on another planet. Dan Goldin of NASA called the discovery 'exciting, even compelling, but not conclusive'.

Finding imprints of primitive life (even if they were only a thousandth the width of a human hair) has led to a stepping-up of plans to explore Mars and calls for an intensive effort to focus on the search for Martian life. US President Bill Clinton recognized the significance of the evidence, giving his verbal support for future missions to Mars dedicated to finding more evidence for Martian life, either fossilized or still extant in Mars' soil.

If the impact that blasted ALH84001 from Mars took place 16 million years ago, then the crater which was produced is likely still to be visible. One possible site has been suggested – an elliptical crater measuring 23 x 10km in the Sinus Sabaeus region, produced when a large meteoroid struck Mars in a glancing blow that sliced a portion of Martian crust into space.

The measured proportion of the light isotope carbon 12, produced from the radioactive decay of carbon within ALH84001, indicates that it was derived from a form of methane (Martian marsh gas), which is made of carbon and hydrogen and is likely to have been produced biologically. Further evidence to support the existence of ancient primitive Martian life has been found in other Martian meteorites. One specimen found in 1979 contains 50 times more organic material than ALH84001. This rock is thought to have formed 180 million years ago (when the Earth was in its Jurassic period) and was ejected from Mars 600,000 years ago.

Now the search for more clues among the few samples of the Martian surface that we have in our possession is developing apace. In a roundabout way, cosmic impact is enabling us to learn a great deal about the make-up of our planetary neighbourhood and has raised hopes that life is not a phenomenon unique to the Earth.

6

WATCHING THE SKIES

OBSERVING METEORS

Can there be anyone upon the Earth who has not been struck by the phosphorescent lights that glide through the sombre night, leaving a brilliant silver or golden track – the luminous, ephemeral trail of a meteor?... sometimes... a shining speck is seen to detach itself... from the starry vault, shooting lightly through the constellations to lose itself in the infinitude of space.... Those bewitching sparks attract our eyes and chain our senses. Fascinating celestial fireflies, their dainty flames dart in every direction through space, sowing the fine dust of their gilded wings upon the fields of heaven. They are born to die; their life is only a breath; yet the impression which they make upon the imagination of mortals is sometimes very profound.

So wrote Camille Flammarion, a nineteenth-century French astronomer and science popularizer noted for the lyrical way in which he described heavenly phenomena (see Bibliography). It is true that most meteor observers apply themselves to their nocturnal business because they gain access to inner peace as they contemplate the canopy of stars above, and for them – more often on mild nights with good levels of meteor activity – the words of Flammarion ring especially true. To be engaged in meteor observation may bring about a strange state of mind, a kind of celestial angling mentality where most of the time thoughts are allowed to wander; yet suddenly, without warning, this meditation must be left at an instant to make a record, a brief note in a book or a few words on audio describing the meteor's appearance.

Although meteor observation may have a degree of romantic appeal, dedicated observers are a hardy breed. It takes a special kind of dedication to be willing and able to spend hundreds of hours each year, sometimes in sub-zero temperatures, notebook or tape recorder in hand, looking at the heavens with the intention of recording the details of each meteoric flash that presents itself through a chosen area of sky. Those who elect to do this know that meteor observing is a valuable scientific pursuit, one of the few branches of amateur astronomy which is still capable of yielding important new information about the smallest members of our local neighbourhood in space.

For all the obvious merits of meteor observation (spiritual as well as temporal), most watchers of the night sky find it difficult to muster the dedication necessary to graduate from being a cursory meteor watcher to joining the ranks of competent meteor observers. Most amateur astronomers would rather spend their time glued to the telescope eyepiece, looking at more static celestial objects at an hour suited to them. We conveniently choose to become avid meteor watchers only when a productive annual shower nears the date of its maximum and there is a guarantee of success (if success is measured by the sheer number of meteors seen)!

Preparation

John Haldane, a prolific British popularizer of science, once wrote that 'the observer of meteors requires a clear sky, a thick coat, a notebook, a knowledge of the constellations, infinite patience, and a tendency to insomnia' (quoted by Stephen Edberg and David Levy – see Bibliography). Perhaps his list should also have mentioned a comfortable observing posture, because whether you are a fair-weather shooting star spotter or a dedicated meteor observer, there is every reason to be as comfortable and relaxed as possible – it is as important as a stable mounting is to the telescopic observer. Being comfortable while 'in the field' improves observational accuracy and efficiency and, most importantly, is vital to the observer's enjoyment of the hobby.

Meteor observation requires a good view of the selected region of the sky, preferably one which allows the sky to be seen down to the local horizon and is uncluttered by foreground trees or buildings. A reclining chair or garden lounger with some head support is infinitely preferable to an upright stool with no back support, and unless you are planning only the briefest of meteor observing sessions, standing up is liable gradually to induce unnecessary levels of physical and mental fatigue. Hour-long observing sessions followed by indoor breaks of ten minutes are ideal for beginners.

No single observer can hope to cover the whole sky, so if you are alone it is best to face the radiant of the expected meteor shower and allow your eyes to wander comfortably over the region of sky in the radiant's vicinity. It is a useful exercise to try to identify the adjacent constellations and brighter stars in the area during the session, so it pays to have done a little pre-observing research. If you spot anything you are not familiar with, make a note of the star or asterism in question and resolve to look into it when you are indoors – if you spend your time consulting your star atlas in the field you are probably going to miss some meteors!

Recording

A hand-held micro-tape recorder will allow you to note everything you see without having to take your eyes off the sky for too long in order to write things down and possibly miss out on meteors in the process (note that most digital message recorders have a very limited recording length and therefore may not be suitable for your purposes). Unless you like to hear the sound of your own voice and are not worried about the exorbitant cost of batteries, try to keep idle chatter to a minimum and stick to a plan of verbal action. An accurate watch (or clock) with a clear digital readout is essential, and it might be worthwhile obtaining one with a luminous face to save time spent fumbling with torches. Some clocks and watches become inaccurate in adverse weather conditions (not necessarily the cheap ones, either), so keep them warm and shielded.

The tape recording should begin with the date and time the session was commenced, your observing location, a summary of the weather conditions (including cloud coverage) and a note of the limiting magnitude (the brightness of the faintest stars visible at that time with the naked eye). Most meteor observers shun moonlit nights because the fainter meteors are drowned by the glare – that is, unless they are compelled to observe because it is a special unmissable occasion such as a shower maximum or a predicted storm. So, if the Moon is visible, note its height above the horizon and its phase.

Remember to keep the tone and volume of your voice on the low side, so as not to set off a cacophony of assorted animal noises in your locality. It has been known for amateur astronomers to find themselves under the scrutiny of the police, alerted to your mysterious nocturnal activities by a neighbour who thinks you are up to no good; while you are busy explaining the intricacies of meteor observation to a disbelieving police sergeant at the local station, you might be missing out on a spectacular display.

Once a meteor is seen, it will usually take a few seconds (more if a magnificent fireball is seen) for you to turn from the heavens, look at the time, press 'record' and state clearly:

Mark... Time [UT]: Xh Xm Xs [to nearest five seconds, if possible]....
Shower/Sporadic.... Meteor/Fireball seen... estimated brightness... colour
[if any]... apparent speed [slow, medium or fast]... length of flight [to
nearest 10 degrees if possible]... unusual features [such as flaring,

fragmentation, exceptional slowness or persistent train]... direction of flight from radiant [if shower member].... End.

If a fireball is seen, then an estimate of its apparent velocity may be made on a scale of 0 to 5, where 0 = stationary, 1 = very slow, 2 = slow, 3 = medium speed, 4 = fast, 5 = very fast. As for its estimated brightness, give a range of possible magnitudes for an individual fireball, say between magnitudes -6 and -10, rather than just one magnitude which will probably be inaccurate, even if you are a veteran meteor observer. If the fireball fragmented in flight, note the number of particles seen and where along the fireball's path they appeared to separate from the main body. A drawing, made as soon as possible after the sighting, is invaluable.

I have found that a cue card with each of these important points written on it is invaluable, just in case one or more is inadvertently neglected. Much the same information can be written by hand, although it will probably take longer to record than a well-rehearsed tape message. Remember to record any periods of increased cloud cover and make sure that you note the times when you took your eyes off the sky to have a short break. Keep written notes tidy by writing them in sequence on lined paper with columns drawn for the various data. If a spectacular event is seen, such as an exploding fireball, you should attempt a drawing, fully annotated and showing the stages in its flight against the background constellations.

After the observing session, write up your taped or written notes neatly in sequence on a report form. It is not advisable to include observations made on two different nights on the same form – to avoid confusion, always fill in a fresh observing form for each individual night of observation. There is something very forlorn about a set of observations that is simply filed away in an individual's observing drawer, never to see the light of day – even train spotters brag about their achievements and compare notes! It is best to be a member of a local or national astronomical society, where you can communicate with like-minded meteor enthusiasts and perhaps arrange to be part of a larger observing group (see Useful Addresses).

Photography

You don't need to be an expert with expensive and sophisticated camera equipment to capture a meteor on film. Any camera that allows its shutter to be left open to take a time exposure can be used. An ordinary camera, loaded with normal 200 ASA colour film and laid flat on its back looking directly overhead on a dark night (away from artificial light sources) is bound to capture hundreds of curved star trails, a small assortment of coloured aircraft tracks and, possibly, one or two satellite trails, given an exposure of around 15 minutes. Longer exposures, however, stand a strong possibility of being washed out by light pollution if the photograph was taken from an urban location. If the night you choose to take your photographs happens to coincide with a meteor shower maximum, then you would be unlucky not to record the trail of at least one meteor on a 15 minute exposure, provided the camera is pointing towards the direction of the shower's radiant.

Most SLR cameras are provided with a general purpose, 50mm focal length lens which may be used at the f2 setting (maximum aperture), and with this the whole of the constellation of Leo may be captured on the image. However, with such a medium-sized field of view, many meteor trails may be only partially captured as they speed out of the image field, especially if the camera is pointing some distance away from the shower radiant, where meteor trails appear longer. A wide-angle lens (say, 28mm) is an advantage, since it will capture a greater area of sky and increase the chances of capturing more meteors and their entire trails. A film speed of 400 ASA will give good results with a reasonably dark sky background, when taken from a light-shielded urban setting using exposures of less than ten minutes' duration. A fish-eye lens – one that gives a very wide field of view (in some cases 180 degrees) – will give complete sky coverage; however, if you are using ordinary film then only the brightest of meteors and fireballs will be captured.

Meteor photography can provide valuable information about the precise location of a meteor against the stellar background. With a 'chopper' device – simply a motorized rotating shutter which obscures the field of view at regular intervals at a predetermined rate – it is possible to calculate how fast the object was travelling against the starry background. The image of a meteor captured in this manner is chopped up into small sections and appears as a dashed line. Because the speed of the shutter is known, the length of the short trails can be measured, enabling the meteor's apparent angular speed to be determined.

It is important to make a careful note of the time of start and finish of each of your photographic exposures. It is recommended that you take a normal photograph of a well-illuminated daytime scene at both the beginning and end of the film, so that high street developers will know exactly where to cut the images – you do not want all your beautiful, hard-won meteor photographs chopped in half by someone who is not aware that those dark exposures with a few specks of light are of value to you.

OBSERVING ARTIFICIAL SATELLITES

To the naked eye, a satellite in a low orbit around the Earth appears as a single white point of light that slowly crosses a part of the sky in a straight line. The light seen is simply sunlight reflecting off the shiny exterior of the satellite – if a satellite does happen to have its own lights, they are certainly not bright enough to be seen with the naked eye from the Earth's surface. So, if you happen to observe an object with blinking coloured lights and/or an erratic path across the heavens, then that object is bound to be an aircraft operating within the Earth's atmosphere.

Satellites can appear to vary in brightness gradually during the observation, perhaps appearing at first as a brilliant object as bright as Venus, fading to a barely visible point of light some moments later and then vanishing altogether. This is an effect produced by the object's passage from full sunlight into the shadow zones that the Earth's globe casts into space. Like the Earth's natural satellite, the Moon, artificial satellites endure these eclipses as they plunge into and out of the Earth's shadow. Most low-orbiting satellites go into shadow eclipse on every single orbit, perhaps as often as once every one-and-a-half hours. Shorter frequency variations in the satellite's observed brightness are caused by the satellite turning end over end in space, reflecting different amounts of light towards the observer as it tumbles, and are best seen when the satellite is illuminated by full sunlight.

Objects in the lowest sustainable Earth orbit fly at a height of around 275km and make one Earth orbit in 90 minutes; when seen, the object will appear to cross the observer's sky in a few minutes. Very low-altitude satellites are likely to be manned spacecraft, supply craft for the Russian Mir, temporarily orbiting booster stages, or spy satellites. At a height of 4,180km, a satellite makes one Earth orbit in three hours. Satellites at this altitude may be specialized science or resources satellites making observations of the Earth from a near-polar orbit.

At a height of 36,000km above the Earth's equator, a satellite makes one orbit in exactly 24 hours, thus matching precisely the Earth's own daily revolution on its axis. Seen from the Earth, the satellite will occupy one stationary point in the sky – such a 'geostationary' satellite is, in effect, hovering effortlessly above one point on the Earth's equator. Geostationary orbits have been chosen for communications satellites for many years – there are at present some 200 operational geostationary satellites clustered at strategic points over the Earth's equator, and the number is growing month by month as the world's communications needs grow. Much further than 1.6 million km from the Earth, an orbiting satellite can be classified more appropriately as having a solar orbit in the Earth's vicinity, for its orbit is at all times concave to the Sun.

Of nearly 7,000 satellites in orbit, most of which are very faint objects down to the tenth magnitude, hundreds can be seen through binoculars, given full illumination and a good position relative to the observer and the Earth's shadow. Observer Gordon Taylor of Cowbeech in Sussex has even reported that a number of geostationary satellites, between the 11th and 13th magnitudes at a distance of around 36,000km, can be seen telescopically.

Less than 400 of the observable Earth-orbiting objects are currently operational satellites; the rest are dead satellites or pieces of debris, including parts of rocket launchers. There are currently around half a dozen regular naked-eye satellites, including the Mir space station (orbiting at an altitude of between 300km and 400km every 90 minutes), the US Seasat (a 12 x 1.5m science satellite launched in June 1978) and the Space Shuttle, visible during its frequent week-long journeys into space. For cursory observers, special satellite happenings – such as the orbital link-up of a Space Shuttle and the Russian space station Mir – are liable to be publicized in various news media and some astronomical magazines, and, provided you make your observation at the right time and look in the right direction, are worth making the effort to observe.

Dedicated satellite observers find it satisfying not only to be able to see a light in the sky and recognize it as a satellite, but also to know the name (or designation) of that satellite, its country of origin, its height and orbital period, and how long the satellite has been orbiting the Earth. For more than four decades, visual observers have made a great contribution to our knowledge of satellite orbits, the effects of drag caused by the upper atmosphere and the space-faring activities of foreign powers.

Gathering Data

When learning to track satellites visually, it is best to begin by practising on brighter satellites whose orbits have already been predicted accurately, thus providing a yardstick by which the observer's accuracy may be gauged. When a satellite is seen, the observer notes its passage across a familiar part of the sky, say between two known bright stars; usually, the smaller the separation between the two stars, the more accurate the positional measurement is going to be. A timing is made with a stopwatch when the satellite crosses an imaginary line drawn between these two reference stars, and an estimation is made as to where the satellite appeared upon that imaginary line. Once an accurate timing and a position have been secured (by plotting on a star map and taking its coordinates), this information can be sent to a centre for observations, where it may be reduced and used as raw data to calculate the satellite's orbital path. With enough dedicated satellite observers, amateur observations are capable of producing impressive results (see Useful Addresses for contacts). Anyone inspired enough to progress to binocular work in tracing the fainter Earth-orbiting objects will find the prospects for further research truly awesome; nowadays, the field is somewhat neglected and is always in need of fresh new observers.

7

FAMOUS FIREBALLS, UFOS AND METEOR MISCELLANY

The sudden, unexpected appearance of a brilliant meteor or a fireball is quite capable of inducing utter astonishment in the most seasoned watchers of the skies. Today, most people don't bother to take the time to view the heavens – especially now that the night skies of many urban areas are bathed in an unattractive orange wash created by the light pollution of millions of sodium streetlights. It is hardly surprising that a city dweller, unused to the true splendour of the heavens, who happens to glimpse a short-lived piece of meteoric action, might imagine that mysterious (even supernatural) events are taking place in the skies. Most people who report having seen a nocturnal UFO have actually seen the flash of a meteor or the blazing descent of a fireball. Their imagined extraterrestrial visitors in spaceships are, in fact, grains of extraterrestrial dust or larger chunks of rock on a one-way trip to the Earth.

SONIC SHOOTING STARS

Some fireballs, travelling faster than the speed of sound as they penetrate through to the lower levels of the atmosphere, can produce sonic booms audible to those on the ground. A sonic boom is a noise like an explosion, caused when a shock wave of compressed air, formed against the leading surface of the meteoroid, reaches the ground at supersonic speeds. Sometimes there can be a lengthy series of noises, like a drawn-out rumble of thunder from a distant storm.

A typical account of a brilliant fireball, including sonic effects and a persistent train, is given by an officer of the USS *Alaska*. The following extract is taken from *Knowledge*, February 1883. The fireball was seen from on deck on the evening of 12 December 1882, a few minutes after sunset:

> All at once a loud rushing noise was heard, like that of a rocket descending from the heavens with intense force and velocity. It proved to be a meteor, and when within ten degrees of the horizon, it exploded with much noise and flame, the fragments streaming down into the ocean like great sparks and sprays of fire. The most wonderful part of the phenomena then followed for at the point of the heavens where the meteor burst there appeared a figure shaped like an immense distaff, all aglow with a bluish light of intense brilliancy.... It kept that form for perhaps two minutes, when it began to lengthen upward, and growing wavy and zigzag in outline, diminished in breadth until it became a fine, faint spiral line, and its upper end dissolving into gathering clouds. It remained for about ten minutes, when it began to fade, and finally disappeared.

Northern Lights, a Fireball and a Bang

The night sky of 8 November 1991 will be remembered by British astronomers as having hosted two remarkable events: a spectacular auroral display, and a brilliant fireball that produced a (probable) sonic boom. As my wife and I were driving through Wales that evening with a schedule to meet, we took the opportunity to pull over for a short while and marvel at the aurora, a wonderful vision of ruby-coloured streamers moving slowly to the rhythm of the electromagnetic forces controlling their manifestation high above the atmosphere. I telephoned my friend Paul Stephens back in the Midlands, alerting him to the phenomenon, and consoled myself with knowing that least somebody, now, would be able to make a complete record of the event and take photographs as the aurora unfolded.

At around 11 o'clock, as we passed through a wooded area on the outskirts of a small village in mid-Wales, the interior of the car became briefly illuminated, and the map I had been trying to read by red torchlight was temporarily bathed in white light. An astonished Tina, who was driving, told me that she had seen an intensely bright ball of light flash low across the sky and apparently drop down beyond the trees. I must admit that thoughts of 'UFO encounter on lonely country lane' did cross my mind, but as reason prevailed, and a full description of the objects's flight path and appearance was obtained, I realized that this had been an impressive fireball.

Meanwhile, 130km west of our location, Paul had acted on my message and was making observations of the aurora. Incredibly, he also observed a brilliant fireball at around the same time as our experience; the chances are that it was one and the same object, observed from different angles. In his comprehensive report on the aurora, Paul stated:

As I prepared to take my first picture at 22.55, the same area of sky [to the northeast] was crossed by a brilliant fireball, of at least magnitude -6, starting near Capella and ending in Ursa Major, intensely white with a blue margin and illuminating its trail momentarily as it sped across some 40 degrees of sky. Soon afterwards there was a loud report [the sonic boom], but I keep an open mind on this remembering that we were only three days past Bonfire Night!

Other Sounds

In addition to conventionally explicable sonic booms, some really bright fireballs produce hissing or crackling sounds as they cross the sky. Meteor observers have known about this strange phenomenon for a long time, but it was only widely accepted after a fireball of magnitude -16 (about ten times brighter than the full Moon) passed over Sydney, Australia, on 7 April 1978. Out of several hundred witnesses known to have observed the great fireball, one-third of them attested to hearing a hissing noise as the object passed over. Two people located indoors heard the hissing noise without having seen the fireball. Physicist Colin Keay of the University of Newcastle in New South Wales looked into the phenomenon of hissing meteors, and came to the conclusion that turbulent ionized gases formed in the fireball's wake can transmit very low frequency (VLF) radio noise, which may be detected by the human ear or felt indirectly as a result of vibrations produced in metallic objects in the observer's vicinity. With the aid of volunteers, Keay demonstrated that VLF noise could be detected by humans, but that there is a wide variation in perception from subject to subject. It was also shown that the VLF emissions of aurorae (from 2kHz to 6kHz) lie within human range, and there is plenty of evidence to demonstrate that aurorae have been accompanied by distinct hissing noises.

SEEING IS BELIEVING

UFO Connections

Before the twentieth century, most people who saw a bright meteor or fireball power its way across the sky were genuinely mystified by what they had witnessed, and their initial interpretation of the sighting would have depended on their familiarity with the night skies, their level of education and their religious beliefs. In those days, no fireball witness would have believed that they had watched the arrival of an alien spacecraft, but they may have imagined that angels or demons were involved up to their necks in the phenomenon. In 1898, H. G. Wells wrote *The War of the Worlds,* in which he chose to have his malevolent extraterrestrials arrive from their home planet Mars within giant meteorites. Since then, an increasing number of people, more familiar with science fiction than science fact, have interpreted fireballs as alien spacecraft.

A typical fireball that prompted a spate of UFO reports occurred in the English Midlands on the evening of 18 February 1971. One observer, a woman driving her car near Alvechurch, reported observing a very bright object 'about the size of the Moon, blue with a white centre'. During the five-second sighting, the object travelled southwest and the astonished woman claimed it had set down in a nearby field. Several independent observers in a variety of locations saw an identical brilliant object at the same time, and the *Birmingham Post* quoted one meteorological expert who claimed that the witnesses had seen a 'shower of meteorites... this is about the right time of year for them'(!). In fact, the object was undoubtedly an impressive fireball which illuminated hundreds of square kilometres of the English Midlands in its fiery passage through the Earth's atmosphere. Although the woman imagined it had landed locally, the fireball may actually have been very far away – so far that it dipped below the distant horizon while still in flight. Moreover, unless fragments of the object were recovered, there is no way of telling whether the meteoroid at the centre of the fireball survived intact all the way down to the ground.

Point Meteors

On 12 May 1971, two people in the West Midlands, England, observed a fiery red ball of light suddenly appear in the northeastern sky. The witnesses claimed the object to be 'six to eight times as big as the Sun' (most people who use this expression really mean 'brighter than' rather than 'bigger than' the Sun or Moon). The object did not move against the sky background and was only visible for a moment. The witnesses were adamant that the object was not a flash of lightning. A similar sighting was reported to me by three people who, on the clear Moonless evening of 4 February 1994, had been startled by the sudden materialization of a stationary, brilliant golden-coloured orb 'about a quarter the size of the Sun' in the northern sky over Dudley, also in the West Midlands. No apparent sound emanated from the object, and the observation was estimated to have lasted 20 seconds (most untrained observers tend to exaggerate the duration of sightings). The observers agreed that the object was clearly defined and appeared solid; however, one observer disagreed that the object was an 'orb', and another added that there may have been a red 'tinge' to it.

It is likely that most of these apparently stationary brilliant objects of short duration are views of ordinary meteors approaching the Earth in flights aimed directly at the observer. Faint examples of these point meteors may actually be observed during every one of the more active annual meteor showers, and they appear at or very close to the shower's radiant. However, bright examples of point meteors represent a proportion of perhaps as little as one ten-thousandth of all the showers' brighter meteors. The chances of directly observing a sporadic (non-shower) point meteor are far less than observing an ordinary meteor, and because the eye's attention is not drawn to their appearance in the same way that a rapidly moving object commands, they deserve to be classed as rather rarely observed celestial phenomena.

From the cosmic bombardment of millions of meteoroids experienced by the Earth during great meteor storms, many point meteors have been observed near the storm radiant. During the Leonid storm of 12 November 1833, it is reported that an observer at Boston counted 650 shooting stars in a quarter of an hour. Large fireballs with luminous trains were also seen, some of which remained visible for several minutes. Even point meteors are said to have been seen, and one in particular was mentioned as having remained for some time in the zenith over Niagara Falls, emitting streams of light.

EXTRAORDINARY FIREBALLS

On the night of 9 February 1913, thousands of observers across North America, from the plains of central Canada through the US, were treated to a display of extraordinary meteors. During a five-minute period at around 02:00 UT, astonished observers saw a shower of many fiery red- and orange-coloured meteors travelling in small clusters, moving slowly through the atmosphere in horizontal flight paths at seemingly low altitude. Some of the brighter meteors were unusually long lived, visible for a minute as they crossed the sky. Since the objects had appeared on the feast day of St Cyril of Alexandria, they were unofficially named the 'Cyrillids'. Despite subsequent watches, no trace of the Cyrillids was ever seen again, although on the following evening some unusual activity was observed over Lake Ontario by several people on the outskirts of Toronto.

One theory maintains that the energies released by a meteoroidal impact on the Moon launched a stream of material into Earth orbit; its subsequent burn-up in the Earth's atmosphere was seen as the Cyrillid shower. A more likely explanation is that a large meteoroid surrounded by a cloud of meteoroidal debris grazed the Earth's atmosphere and entered into a very low temporary orbit around the Earth – so low that atmospheric friction soon caused the meteoroids to fragment and burn up. It is possible that many of the objects were substantial enough to make it to the ground, but searches failed to discover any meteoritic material that could be attributed to the Cyrillids.

Perhaps the most famous example of a fireball being widely observed and mistaken for a UFO by many who saw it happened on 10 August 1972, when a spectacular teardrop-shaped fireball appeared in the afternoon skies over the Rocky Mountains in the US. Moving slowly and gracefully (at least to the eyes of ground-based observers), the brilliant object travelled in a straight path southwards over Montana, Wyoming and Utah, leaving a narrow shining trail in the sky. The object covered a distance of 1,500km in 100 seconds, giving it the phenomenal speed of 54,000km per hour and placing it in the lower half of the speed range for most incoming meteoroids. Unlike many other fireballs, this partic- ular object did not explode high in the atmosphere or make its way down to the ground. Instead, the meteoroid approached the Earth at such a narrow angle that it simply bounced off the atmosphere and went back out into interplanetary space, in the same manner as a stone skipping off the surface of a pond.

A US Air Force satellite making infrared observations of the Earth also captured the object on its sensors and recorded the tremendous tem- perature of the meteoroid as it heated with atmospheric friction. Com- bining the ground-based and orbital observations, it has been calculated that, depending on its composition, the meteoroid was originally around 3m to 14m in diameter (even as large as 80m, if some calcula- tions are to be taken seriously) and that it had been in a highly elliptical orbit around the Sun before its brush with the Earth. Its closest approach to the Earth took place over Montana when only 60km high – perilously close to what could have been a local disaster. The meteoroid, having now been gravitationally deflected into an even more elongated orbit, may return to the Earth at some time in the distant future. According to calculations made by astronomer Zdenek Ceplecha, the object came back to the Earth's vicinity in July or August 1997 – were it to hit our planet, then it would impact with an explosive energy equivalent to that of a small nuclear bomb.

On 3 October 1996, a meteoroid as big as a saloon car approached the atmosphere from a shallow angle over the southwestern US, and its break-up and fragmentation was observed. Dr Boslaugh of the University of California described the object as 'a deep red fireball with multi- coloured sparks whirling off its back'. Others, from San Francisco to Mexico, saw a green fireball (possibly a colour contrast after-effect). At a height of about 40km above the town of Artesia, New Mexico, the meteoroid broke up, showering some meteoritic debris beneath it. What makes this event so unusual is that the largest chunk somehow managed to bounce back into space. The preheated meteoroid made an orbit of the Earth and re-entered over the western seaboard of the US, producing audible sonic effects and a dazzling fireball that was widely observed. Excited meteorite collectors offered cash rewards for any recovered samples of the object, with a hefty bonus promised for the first 4oz (113g) piece.

Strange Green Fireballs

One of the most curious episodes in the history of meteor observation took place during late 1948 and throughout 1949 when, on many occasions, brilliant green fireballs were seen in the skies of New Mexico

The objects were usually observed to be much slower than ordinary meteors and crossed the sky in horizontal trajectories rather than appearing to radiate from a single point in the sky. Sometimes the green fireballs terminated with a dazzling explosion, although sonic effects were rarely heard, nor were any definite meteoritic fragments ever found on the ground.

New Mexico was then (and still is) a militarily sensitive region, with numerous US Air Force bases, an assortment of missile and space science research facilities, and nuclear energy plants. The green fireballs were an aerial phenomenon that science could not immediately explain, and because these strange objects regularly overflew these sites they were considered a potential threat to national security. The Air Force invited astronomer Dr Lincoln LaPaz of the University of New Mexico, an internationally renowned expert on meteors, to assist them in their investigations.

Like Jean Baptiste Biot, who investigated the L'Aigle meteorite fall of April 1803, Lincoln LaPaz was initially a sceptic, although he patiently went about his interviews with witnesses and plotted the observed paths of the green fireballs. In December 1948, LaPaz was left in no doubt that the green fireballs actually existed when he and his wife observed a green fireball for themselves. The scientist described his sighting (quoted by Timothy Good – see Bibliography):

This fireball appeared in full intensity instantly – there was no increase in light.... Its colour, estimated to be somewhere around the wavelength 5,200 Ångstroms, was a green hue, such as I had never observed in meteor falls before. The path was as nearly horizontal as one could determine by visual observation.... Just before the end... the green fireball broke into fragments, still bright green.

Mrs LaPaz was later persuaded to do a painting of the object from memory – the work prompted sufficient interest to be printed in full colour in the pages of *Life* magazine.

Colonel Poland of US Army Intelligence was so concerned about the green fireballs that his confidential memo sent to the Pentagon Director of Intelligence, General Staff, Washington in January 1949 reads:

Agencies in New Mexico are greatly concerned over these phenomena. They are of the opinion that some foreign power is making 'sensing shots' with some super stratosphere device designed to be self-disintegrating.... Another theory advanced as possibly acceptable lies in the belief that the phenomena are the result of radiological warfare experiments by a foreign power, further, that the rays may be lethal or might be attributed to the cause of the plane crashes that have occurred recently.

Still another belief... is that it is highly probable that the United States
may be carrying on some top-secret experiments.... It is felt that these
incidents [the green fireballs] are of such great importance, especially
as they are occurring in the vicinity of sensitive installations, that a
scientific board be sent to this locality to study the situation with a
view of arriving at a solution of this extraordinary phenomena with
the least practicable delay.

In February 1949, just over a month after the excited colonel's memo,
the US Air Force invited a select group of military officers, scientists and
astronomers (including LaPaz and nuclear physicist Dr Edward Teller) to
a secret green fireball conference at Los Alamos, New Mexico, in order to
get to grips with the problem. One thing became clear – the green fireballs did not resemble any known aerial or atmospheric phenomenon,
and their nature remained open to speculation. Edward Teller was of the
opinion that the phenomenon represented an ill-understood process
involving high-energy plasmas or coherent electrical discharges in the
atmosphere, but LaPaz was certain that the green fireballs were neither
meteors nor any known natural phenomenon, and concluded that they
were artificial objects powered by an unknown source and going about
their nocturnal hauntings with an equally unknown *modus operandi*.

To investigate the green fireballs and other nocturnal phenomena
further, Project Twinkle was instigated by the US Air Force (under the
Cambridge Research Laboratory, Massachusetts) in September 1949,
headed by LaPaz. It was intended that the project would obtain firm
data on the fireballs – measurements of their altitude, diameter, velocity
and spectrum – using three kinetheodolite stations in New Mexico.
However, only one station became operational, and that provided no
useful data. In December 1951, Project Twinkle was deemed a failure and
further research was abandoned.

With their preference for a single geographical location, slow apparent speeds, extreme brilliance, vivid green colour and unprecedented
horizontal trajectory, combined with the lack of material evidence found
on the ground following their flight, is it possible that the green fireballs
were a very unusual kind of meteor?

Perhaps the most difficult thing to explain is the fireballs' appearance
in the skies over a single area of the globe over a period of many
months. Any stream of objects arriving from the depths of space might
hit the Earth at the same latitude, just as the fragments of Comet Shoemaker-Levy 9 hit Jupiter's south temperate region during that famous
week in 1994. In Jupiter's case, the planet's rapid revolution on its own
axis in under ten hours meant that the impact sites were distributed at
various longitudes around the planet. The Earth is a moving target that
revolves on its axis at an equatorial speed of 1,670km per hour and
moves around the Sun at an average orbital velocity of 30km per second.

In the most favourable circumstances, the green fireballs of the 1948–9 season might reasonably have been expected to be observed at many other locations around the world that lay on the same terrestrial latitude. But a glance at a world map might give some idea as to why the green fireballs were not widely seen – more than half the globe at the latitude of New Mexico is covered by the vast expanses of the Atlantic and Pacific Oceans, and most of the land at that latitude comprises the Asian mountain ranges and sparsely populated desert. The only populated region on the same line is Japan, but that country presents a very small area in comparison to the US. Interestingly, however, the Japanese had their own wave of alleged UFO sightings in the 'Far East flying saucer flap' of 1958!

In 1960, Donald Robey of Convair Astronautics wrote a paper in which it was hypothesized that the green fireballs were small, icy cometoids a few metres in diameter, travelling at a low velocity. Robey speculated that these microcomets were composed chiefly of frozen nitrogen, a substance which is known to give off a distinct green glow when heated to certain temperatures. Robey explained that if the cometoids had been frictionally heated by their passage through the Earth's atmosphere and externally melted to assume spherical shapes, then gas jets emitted by the revolving sphere might produce a force of lift sufficient to keep the object on a peculiarly non-meteoritic horizontal trajectory or even enable it to hover motionless in the sky! Robey calculated that a spherical nitrogen-ice cometoid some 3m across might, at a distance of 32m, appear as bright as the Sun, with most of the observed light being given off in green wavelengths. Observed from a distance of 700m, the cometoid could attain a brightness equivalent to the full Moon.

Robey mentions an observation of a very unusual green fireball, complete with apparent gas jets, that was witnessed by several reputable persons, including newspaper editor Brian Coyne, Trooper Dick Hadsall and members of the Arkansas City Police Department. The incident took place on 19 July 1956 in Arkansas City (1,000km east of New Mexico) and was reported in the San Diego *Evening Tribune*:

> The Kansas State Highway Patrol today said 'a ball of fire travelling east at a high rate of speed had been picked up by radar at the Hutchinson (Kansas) Naval Air Station.' City Editor Brian Coyne, of the Arkansas *City Daily Traveller* said 'a brilliantly lighted, tear-shaped object with prongs or streams of light spraying downward was sighted shortly after midnight. A second object was sighted around 1am. The prongs or streams of bright light also were observed first as directed towards the Earth and then extending from the sides of the object.' He also described the head of the object as being green in colour or 'bluish green'.

Green fireballs have been observed in many other locations around the world during the last few decades. A typical example took place on 14 June 1974, when an exceptionally luminous, slow-moving 'green blob' traversed the sky over Weoley Castle in Birmingham, England. Eight people observed the object during a two-minute sighting (probably an unintentional exaggeration of the timescale), as it changed colour from dazzling white, when first seen, to a vivid green hue. The object was reported to W. E. Marsh, a local amateur astronomer who, it seems, over-reacted somewhat by suggesting that the witnesses immediately contact the Ministry of Defence with full details of their sighting.

In May 1997, Dr Louis A. Frank of the University of Iowa announced startling new evidence that the Earth is under a continuous bombard-ment by small cometoids – 'house-sized' lumps of ice each weighing between 20 and 40 tonnes. Cameras on board NASA's Polar satellite (launched in 1996), which looks down on the Earth from a vantage point high above the north pole, have been recording the vaporization of between 5 and 30 of these microcomets *each minute*! Fortunately, most of these objects are too light and volatile to hit the ground, burn-ing up completely at heights greater than 1,000km.

The objects were first suspected in 1982, when careful analysis of satel-lite photographs uncovered numerous small, dark spots in the Earth's atmosphere – at first these spots were written off as flaws in photography and image processing. However, the Polar satellite has shown that when each microcomet vaporizes, the cloud of water vapour released into the atmosphere temporarily blocks off ultraviolet rays coming from the Earth's surface, giving rise to dark spots visible in ultraviolet images. Astronomers are now coming to terms with this newly announced form of terrestrial bombardment – a new angle on the threat our planet faces from space – in much the same way as science's reaction following the l'Aigle meteorite fall in 1804, which finally led to the acceptance that stones do indeed hit the Earth from interplanetary depths.

MAN-MADE THREATS FROM OUTER SPACE

As if nature doesn't give humanity enough problems to worry about, the latter half of the twentieth century has seen the advent of a man-made menace from above whose odds on causing physical devastation are far greater than the arrival of any 'doomsday' comet or asteroid. War rockets have been developed, ranging from weapons systems capable of lobbing small bombs from one country to another via high-ballistic hops, to powerful intercontinental ballistic missiles (ICBMs) that are able to de-liver total nuclear destruction to any part of the globe within minutes.

Hitler's V-2 ('Vengeance Weapon Number 2') rocket was developed in secret by Nazi scientists at Peenemunde, Germany, during World War II. The rocket, a classic pointed missile shape some 14m tall, weighed nearly 12.9 tonnes on take-off, with a warhead containing 750kg of Amatol high explosive. The V-2's thrust amounted to 25 tonnes – not nearly enough to lift it into an orbit, but sufficient to blast it into a sub-orbital hop with a maximum range of 322km. The weapon's guidance system was not very sophisticated, and it was impossible to direct it towards specific military targets such as buildings or airfields.

The V-2 began to make its indiscriminate, highly destructive presence known to the population of liberated Europe, especially southeastern England, during the closing months of the war. Over 4,000 V-2s were fired in anger, with more than 1,000 of them landing in the London region. The V-2 approached its target at a high velocity from directly above, and the Allies were utterly defenceless against it. The V-2 was the first man-made threat from space.

Nuclear Nightmares

It is fortunate that the Nazis failed to develop the atomic bomb before the end of the war. With such a weapon at Hitler's command, Britain would probably have been forced to surrender unconditionally. A super two-staged V-2 descendant (a weapon which was actually in the final stages of planning at the end of the war) could have reached the US with ease, carrying with it a small nuclear warhead capable of destroying the centre of Washington or disintegrating the skyscrapers of Manhattan.

During World War II, the US successfully developed the atom bomb and used the weapon to destroy the Japanese cities of Hiroshima and Nagasaki in early August 1945. Each of these terrible weapons exploded with an energy equivalent to 20,000 tonnes of TNT. Although it is outside the remit of a book such as this to describe the intricacies of the workings of atomic and hydrogen bombs, or their long-term harmful radiation effects, their immediate destructive power may be briefly described.

On explosion, 'normal' bombs throw out material at high speed and demolish man-made structures, producing their devastation by direct or indirect blast effects on their surroundings. Atom bombs and hydrogen bombs explode and set up a high-pressure shock wave that quickly expands from the point of explosion. When exploded in the air, the nuclear shock wave is called a blast wave, because it produces superheated, high-speed, hurricane-force (plus) winds. The amount of damage done by the explosion depends on the bomb's TNT equivalent, the height at which it explodes (the so-called burst height), and the distance from point zero (the point on the ground directly underneath the explosion).

I shall never forget my own father's description of the atomic-bomb test detonation he witnessed at close quarters while serving with the RAF in Australia during the 1950s. One of thousands of unwitting human guinea pigs, my father sat on the ground, unprotected save for cotton slacks and a light shirt, with his back to the biggest atom bomb Britain has ever exploded. Although the bomb was several kilometres away, it was easily visible to the naked eye, suspended beneath a balloon in the middle of the desert. Moments before detonation, a calm voice broadcast over a loudspeaker instructed the airmen to crouch down and place their hands over their eyes. The flash of the explosion was bright enough to be seen through his palms and my father reported being able to discern his metacarpal bones in the glow. Moments later a hot blast swept through the base, and shortly afterwards, as the temporary sandstorm subsided, the airmen were permitted to turn about to witness the mushroom cloud as it lifted many kilometres above the desert landscape.

A 10-kilotonne nuclear bomb (equivalent to 10,000 tonnes of TNT, or half the destructive power of the Hiroshima bomb), exploded at its optimum damage-producing burst height, will severely damage residential buildings to a distance up to 2km from ground zero, producing hurricane-force damage further out. A 10 megatonne bomb (10 million tonnes of TNT) will cause severe, near-instantaneous damage as far as 20km from point zero.

In addition to the blast damage, the extremely high temperatures produced in a nuclear inferno give rise to short-lived, massive glowing fireballs in the atmosphere. A 10 kilotonne bomb will create a fireball up to 300m in diameter, while that produced within a 10 megatonne explosion may be as large as 5km across. The tremendous heat radiated from these temporary fireballs is capable of producing flash burns on bare, unprotected skin up to distances of 30km from a 10 megatonne bomb. Fires are liable to spring up around the explosion site and spread with all the untameable ferocity of forest fires. Firestorms, like the one that swept through the old German town of Dresden after the massive allied bombing raid of February 1945, are capable of doing the worst immediate damage to life and property. Such effects, thankfully, have not been seen since the nuclear attack on Japan in 1945.

Many of the same effects might be produced by the atmospheric explosion of a small cometary nucleus; the Tunguska Event of 1908 (see page 75) is now thought likely to have been just such a cataclysmic cosmic assault. Large meteoroids or small asteroids might impact with the same explosive energy as a nuclear weapon, but because their explosion (if one does hit the ground) will take place deep within the Earth's crust – if this is any kind of reassurance – the immediate destruction above ground will be far less than the explosion of the same mass of nuclear megatonnage in the atmosphere.

In the post-war era, the world's superpowers devoted a substantial proportion of their resources to building bigger, more potentially destructive nuclear weapons and more efficient rocket delivery systems. For the sake of world peace (or in the name of peace, at any rate), it was decided to use nuclear weapons as a deterrent against going to war with each other or invading each other's allies. There would be no victor in a full-scale nuclear war – those whose fingers were poised over the big red button were in no doubt that destruction would be swift, total and mutually assured.

The long-term effects would last for many years beyond any such insane nuclear exchange. In December 1983, a group of scientists published a paper outlining the 'nuclear winter' theory, in which it was claimed that a nuclear war conducted with just half the nuclear weapons in the American and Russian arsenals would throw so much dust into the Earth's atmosphere that sunlight would be unable to reach large parts of the surface properly for many months (more so in the northern hemisphere, where most of the explosions would take place). A great proportion of plant and animal life would perish in a dark, freezing climate until the dust dispersed and fell back to Earth. The precious filter for harmful solar ultraviolet radiation, the ozone layer, would also be damaged, compounding the problems experienced by our planet's flora and fauna below. Life on Earth is so well established that no nuclear war could destroy it, and as humans (some of them, at least) are so intelligent and adaptable, it is difficult to imagine an Earth without *Homo sapiens*. What would certainly happen is the sudden end of civilization as we know it, an overhaul that would eliminate currently recognized governments, politics and national boundaries.

In 1953, the British rocket expert P. E. Cleator outlined an appalling vision of 'swarms of radio-controlled orbital rockets, loaded with atom bombs, endlessly circling overhead, unseen and unheard, ready to hurtle down to Earth at the touch of a switch' (see Bibliography). Thankfully, Cleator's pessimistic cold war nightmare remained in the realms of shocking fantasy. None of the space-faring powers have ever orbited nuclear weapons and placed the things on stand-by with the potential of a first lethal strike capability – let's pray that such cosmic swords of Damocles are never, ever, forged.

Artificial Fireballs

Since 1957, more than 10,000 artificial orbiting objects have re-entered the Earth's atmosphere and burned up. It is estimated that around 500 trackable artificial objects now re-enter the atmosphere from low Earth orbit each year; most of these objects are pieces of space junk, bits of booster and assorted debris, and only a small proportion were ever fully operational satellites.

The blazing re-entry of a spacecraft can be a sight dramatic enough to cause considerable alarm, and even to prompt a spate of UFO reports from those witnesses who are ignorant of the facts. One of the first widely documented cases of such widespread consternation took place on the evening of 7 March 1960, when the American satellite Discoverer

VIII burned up in the atmosphere over the eastern US. A similar flurry of UFO reports came in after midnight on 8 April 1964, when the re-entry of Russia's Sputnik 2 was observed in the skies over Connecticut all the way south to Guyana (then British Guiana) in South America. On this occasion, some highly imaginative observers had been in no doubt that they had seen the arrival of a giant alien 'mother ship' from which had emerged a squadron of smaller scout vessels. One Virgin Islander claimed to have seen a bright green globe being followed by a cigar-shaped flaming craft!

The Russian satellite Cosmos 954, launched in September 1977, was equipped with a small (non-explosive) nuclear power unit, like many of its predecessors in the Cosmos series. However, an onboard malfunction caused the satellite's orbit to decay suddenly. When faced with a number of official requests by concerned western governments, Russia – some-what grudgingly, in a manner befitting the cold war era – made available information about the nature of the satellite's power source and pre-dicted that the craft would fall somewhere along the great circle path between north of Hawaii to the east of Africa. On 24 January 1978, the craft re-entered the Earth's atmosphere over northern Canada, halfway between the predicted points of re-entry. In the days before the satellite's re-entry, the world's news media were quick to raise anxieties – many of them exaggerated or unfounded – about the wisdom of powering sat-ellites in this manner, and criticized space scientists for not being able to predict exactly where satellites would come down! Anyone reading the stories could be forgiven for thinking that Russia was orbiting potential atomic bombs with total disregard for the safety of the people of the Earth.

A year later, in January 1979, a Soyuz ferry craft was launched by the Russians from Tyuratam. Days later, parts of the low-orbiting space launcher (not thought to be nuclear powered) re-entered the atmos-phere, burned up and landed near the southeast coast of England. Dur-ing a routine early-morning inspection of his domain on 3 February, the greenkeeper of Eastbourne golf course in Sussex came across a 45cm-diameter, 25cm-deep hole in the ground. Thinking at first that it could be a rabbit burrow, the keeper inspected the hole closely and found a buried lump of aluminium alloy some 15cm across and weighing over 1kg. The fragment, initially feared to be radioactive, was taken into cus-tody by the Eastbourne constabulary and tested by scientists from the Appleton Laboratories in Slough, Berkshire, and Ministry of Defence officials.

Skylab's Descent

The re-entry of the US space station Skylab in 1979 was the most anti-cipated atmospheric burn-up in history. Skylab, the first – and so far, only – American manned space station, was launched atop a Saturn V rocket in May 1973. The upper stage of the Saturn launch vehicle remained in orbit for 20 months, finally plummeting into the Atlantic Ocean in January 1975.

Skylab was manned by three astronaut crews from May 1973 to February 1974. The final crew spent a combined total of 367 hours (10 per cent of their total experiment time) on observing Comet Kohoutek with special equipment, taking advantage of their clear above-Earth environment to make observations that were difficult (and in some wavelengths, impossible) to carry out from the ground. The total cost of the Skylab project amounted to nearly $2.5 billion. With such a high investment in mind, NASA had hoped to return to Skylab, boost it to a higher orbit, refurbish the station and supply fresh crews ferried aboard the new Space Shuttle at some time in 1981 or 1982.

However, increased levels of solar activity served to expand Earth's tenuous upper atmosphere and this created increased levels of drag on the craft. Minute though the deceleration was in 'real-time', the cumulative effects were devastating. Skylab's orbit deteriorated faster than anyone had predicted, and on 11 July 1979, during orbit number 34,981, the massive craft – 26m long and weighing 90 tonnes – plunged into the atmosphere over the Indian Ocean. Pieces of Skylab survived all the way down to the Earth, distributed over an area 160 x 4,000km, but many plopped into the sea off Australia's west coast. Ranchers in sparsely populated Western Australia found their estates strewn with hundreds of melted, twisted metallic fragments. Recovered pieces of Skylab were considered valuable; most meteorite collectors still long for a specimen, even though this material has little astronomical significance. Despite NASA's plea for the return of any bits of Skylab that were found, many were sold by finders to interested private buyers. A number of large pieces were found 12km south of Rawlinna, some distance inland from the south coast of Western Australia, and sold to a Hong Kong syndicate. The largest recovered fragment of Skylab was a hefty airlock shroud weighing over 2.3 tonnes. Skylab still holds the record as the largest single manned spacecraft ever to be placed into – and to fall out of – Earth orbit.

In 1991, the Russian space station Salyut 7, unoccupied since 1986, met with a similar set of circumstances when, despite its having already been boosted to a higher orbit to minimize atmospheric drag, increased levels of solar activity accelerated the station's demise. Salyut 7 re-entered the atmosphere over the South Atlantic and parts of the vehicle fell on Argentina.

Riding in a Fireball

The heat levels involved near the exterior surface of a spacecraft during its re-entry into the Earth's atmosphere can exceed those of the Sun's surface (the solar photosphere is 5,500°C). A shock wave just forward of the spacecraft compresses the air to incredible levels, and when gases are compressed they heat up. An ultra-heated shell of plasma is formed, which temporarily blocks radio communication between the crew and ground control.

Before the Shuttle era, all manned spacecraft were covered with an ablative substance that burned off, carrying with it the high heat of re-entry as the outer layers peeled away. The US Space Shuttle is a winged spacecraft covered with tough, heat-resistant, insulating ceramic tiles, which are permanently glued on, rather than being a one-off ablating substance designed to withstand just one re-entry. During the Shuttle's re-entry, temperatures of 1,370°C are reached on its underside and leading wing edges. The craft descends gently through the atmosphere and glides, unpowered, down to its landing site. The crew inside feel nothing of the heat of the superheated plasma about them, although they can see it shine brightly though the Shuttle's toughened, heat-resistant glass windows.

Cosmonaut Yuri Gagarin was the first human to ride within a fireball from outer space as his capsule, Vostok 1, plunged through the atmosphere after completing the historic first manned orbital flight on 12 April 1961. The craft's retro-rockets fired over Africa and it began an 8,000km downward curve through progressively thicker parts of the Earth's atmosphere. During the descent, parts of the outer shell of the spherical capsule hit temperatures up to 10,000°C, but environmental systems inside the vehicle maintained an air temperature of a comfortable 20°C within the cabin. Quoted by Peter Fairley (see Bibliography), Gagarin reported:

> Through the porthole I saw the frightening crimson reflection of flames raging all round the ship. But in the cabin it was still 20°C despite the fact that the ship was hurtling toward Earth like a ball of fire.

Although he had the option (a perfectly honourable one, too) to eject from the Vostok capsule during the final minutes of the mission and float to the ground independently by parachute, Gagarin elected to remain within the craft all the way down to its soft landing in a field near the town of Saratov, in the Volga basin north of the Caspian Sea. A witness to Gagarin's landing, a local woman named Anya Takhtarova, was the first person to approach the scene. In his dayglow orange pressure suit, like a humanoid alien visitor in a science fiction movie, Gagarin walked away from the hot, smoking craft. Takhtarova nervously asked the smiling astronaut 'Did you really come from outer space?'

The US chose a different shape for the Mercury capsule, America's first manned spacecraft. Designers had looked at a variety of capsule shapes and had asked a dozen aerospace companies to come up with their own suggestions. The main factors to be taken into consideration were capsule weight; how effective its insulation was against atmospheric friction; how forgiving it was to errors in navigation and what margin of error there was in angle of approach; how controllable the craft was during re-entry; and keeping the G-forces of rapid deceleration within safe limits. In the end, after thousands of design proposals and many hours of scale model testings in wind-tunnels, NASA chose a blunt cone with a rounded ablative heat shield.

Friendship 7, piloted by John Glenn, was the first US manned orbital flight. During the craft's three-orbit mission on 20 February 1962, a number of surprises were sprung on both the astronaut and mission control. One hour after insertion into orbit, as the craft entered sunrise while passing over the Californian coast, Glenn was astonished to observe thousands of luminous particles floating outside the porthole. The astronaut wondered what this debris could be, and had the presence of mind to activate the onboard camera to photograph the phenomenon. Lovely though they appeared to be, the particles observed by Glenn are the first record of man-made space debris interfering (superficially, in this case) with a manned spaceflight. The particles are believed to have been tiny fragments of ice liberated from the craft's exterior during orbit. Quoted by David Baker (see Bibliography), Glenn reported:

> This is Friendship 7. I'll try to describe what I'm in here. I am in a big mass of some very small particles that are brilliantly lit up like they're luminescent. I never saw anything like it! They're round, a little... they're coming by the capsule and they look like little stars. A whole shower of them coming by... they're going at the same speed I am, approximately. They do have a different motion, though, from me because they swirl around the capsule and then depart back the way I am looking. There are literally thousands of them!

During Glenn's final orbit, a signal telemetered to the ground indicated that the craft's heat shield had unlocked prematurely. The shield was meant to unlock automatically during the last few minutes of the flight after the parachutes had deployed, allowing an air-filled landing bag to drop down from the craft in order to cushion its impact on the sea. Attached to the craft's base was a retro-rocket package held in position by three metal straps; in orbit, the retro-rockets were held firmly by the straps, and jettisoned after the rockets were fired to effect re-entry. Mission control knew that if the signal was correct, and Glenn's heat shield had disengaged in orbit, then the only option giving any chance of survival was to keep the retro package on during the whole of re-entry, hoping that the metal straps would hold the heat shield close to the craft's base for as long as possible during the fiery descent.

Glenn immediately suspected that there were problems with the heat shield, and he later admitted to having felt annoyed that mission control did not keep him fully informed as to the true (or perceived) status of his craft. When there was a loud jolt in the early phase of re-entry, Glenn thought the pack had fallen away (it was just the noise of one of the straps breaking). Through the porthole, the astronaut saw flaming fragments up to 20cm in diameter shoot past – actually parts of the retro package – but Glenn was convinced that this was the disintegration of his vital heat shield. During re-entry, Glenn suspected that he could feel the high heat (all 1,650°C of it) of atmospheric burn-up on his back and that the craft was gradually melting around him. He was, however, an ultra-cool professional, who before the flight had been tested physically, psychologically and intellectually. He therefore did his best to adhere to a full flight plan, and showed little sign of panic even in this apparently dire situation. As soon as Friendship 7's drogue parachute was released at more than 8,000m, it dawned upon Glenn that he had a good chance of survival; the relief in his voice is evident.

In his excellent autobiography *Carrying the Fire* (see Bibliography), astronaut Michael Collins vividly describes the re-entry phases of his two space missions – that of Gemini 10 and Apollo 11 – during which he, along with his fellow astronauts, was a passenger riding within a real fireball. The Gemini 10 capsule re-entered the Earth's atmosphere at a speed of 7.6km per second on 21 July 1966:

> As we enter the upper atmosphere, a sheath of ionized gas will surround our spacecraft and create a barrier which our radio signals cannot penetrate, giving us five minutes of 'blackout', or radio silence... now we pass an altitude of 400,000 feet [122km], the height generally considered to be the top of the atmosphere... we are upside down and backward, presenting the blunt heat shield behind us to the increasing friction of the thickening atmosphere. The heat shield, made of a fibreglass, honeycomb structure filled with a silicone material, is designed to dissipate heat by the process of ablation, in which the heat shield will partially erode, literally evaporating and carrying with it the heat of atmospheric friction. It's not long before I can tell that the heat shield is doing its job. We are developing a tail. Tenuous at first, then thicker and more startling, it glows brightly, a red and yellowish gassy tail curving off into the lightening sky.... Occasionally a small chunk of heat shield breaks loose and adds sparkle to the halo.... 'Boy, we're going to hit [the sea] like a ton of bricks!' Amazingly we don't, but plop gently into the Atlantic. We must have caught the lip of a descending wave.

Collins' description of the re-entry of the Apollo command module Columbia, which took place three years and three days after his Gemini 10 experience, is equally dramatic. The spacecraft approached the Earth from the Moon at a velocity of 11km per second, nearly half as fast as Gemini 10. An Apollo command module coming from the Moon needed to approach the Earth's atmosphere at just the right angle to effect a safe re-entry. Too steep, and the craft, travelling at more than 36,000km per hour (and accelerating) would burn up completely; too shallow, and it would bounce off the atmosphere and head off into its own orbit around the Sun, without a hope of return. To keep the strong G-forces to a minimum and allow heat to dissipate adequately from the heat shield, the Apollo command module descended to 55km before lifting to a height of 64km, after which the craft started on its final descent path. From first touching the Earth's atmosphere at a height of 122km down to splashdown, the Apollo craft carved a path through the atmosphere around 2,500km long.

Coming in BEF (blunt end forward), Columbia hit the plane of the atmosphere at an angle of 6.5 degrees and headed towards a patch of empty Pacific Ocean about 130km southwest of the Hawaiian Islands. Indeed, the 2,700°C fireball produced by Columbia's descent over the Pacific was observed from the Earth below, appearing virtually indistinguishable from the burn-up of a genuine meteoroid, and it was photographed by a radar-controlled camera carried aboard a US Air Force Boeing 707 jet. Collins describes the unique experience from within the fireball:

> ... the deceleration is heralded by... the beginnings of a spectacular visual display out the window. We are in the centre of a sheath of protoplasm, trailing a comet's tail of ionized particles and ablative material as we plummet obliquely through the upper atmosphere... a wispy tunnel of colours: subtle lavenders, light blue-greens, little touches of violet, surrounding a central core of orange-yellow, and surrounded... by the black void.... The view out the window is breathtaking. The intensity of illumination has increased dramatically, flooding the cockpit with white light of startling purity... we seem to be in the centre of a gigantic electric lightbulb, a million watts' worth at least.... Our fireball must be spectacular as seen from the pre-dawn murk below, but the rainbow hues are ours alone, too subtle to penetrate the thick lower atmosphere.

Orbiting Debris

Interplanetary space is full of natural debris, and since the launch of Sputnik in 1957 the meteoroids and asteroids have been joined by a growing number of man-made objects. In 1978, two decades after Sputnik, there were 4,500 objects known to be in orbit around the Earth. At the end of the twentieth century there are at least a million individual man-made artefacts heavier than 1kg outside the Earth's atmosphere. Some of these are safely anchored to the Moon, Mars and Venus. Some are coasting around the solar system in their own distant orbits around the Sun and planets, and a few of them are shooting off into deep interstellar space, never to be seen again. But the bulk are satellites, used rocket stages and countless fragments of space hardware that reside in orbit around the Earth at altitudes between 200km and 5,000km. These objects are the true 'vermin of the skies', and the vast majority of them cannot be detected with even the most sophisticated radar systems at our disposal.

In April 1997, a Pegasus rocket launched into low Earth orbit one of the oddest payloads in space history – a 1kg package attached to the launcher's third stage containing the ashes of 22 cremated people, including Star Trek creator Gene Rodenberry, the Apollo engineer Kraft Ehricke and space colony visionary Gerard O'Neill. The original plan of the space funeral company Celestis was to allow each lipstick-sized container to float freely in Earth orbit, but space has become so crowded with man-made junk that it was agreed to keep the containers attached to the rocket stage, which will eventually re-enter the atmosphere early next century after several hundred thousand orbits.

If you go outside on any clear, dark evening and look up at the sky for any length of time, you will observe the occasional flash of a meteor burning up in the Earth's atmosphere at an average height of 80km. You are also bound to see one or two orbiting satellites make their way across the sky; orbital mechanics dictate that the faster the satellite is observed to move, the closer it is to the Earth. The lowest maintainable low Earth orbit is one of about 160km in height – much lower, and the satellite will be skimming the outer atmosphere and will experience considerable drag, which will slow it down and eventually cause it to fall from orbit.

With their arrays of shiny solar panels, the largest satellites are usually highly reflective, and those in low Earth orbit can be seen with ease from the Earth's surface when the illumination conditions are right. With the sophisticated ground-based radar of NORAD (North American Aerospace Defense Command) – originally designed to monitor foreign space activities and to warn of incoming missiles – objects as small as a tennis ball in low Earth orbit can be tracked. While NORAD keeps tabs on over 7,000 objects in Earth orbit, this detectable material represents only the tip of the iceberg, because for every known piece of debris there are many thousands of smaller objects.

The future of manned space exploration will depend largely on how safely humans are able to live and work in that hostile environment, and the hazards represented by both natural and artificial objects is one major factor that determines just how safe space is. It is a sobering thought that a meteoroid as small as a thumbnail or a metallic fragment of space junk, travelling at an average meteoroid velocity, can hit the hull of a spacecraft with all the energy of a high-velocity rifle bullet. The impact of faster meteoroids will have the same effect as a rocket grenade, and lethal consequences will ensue unless the spacecraft is shielded or there are onboard safety measures to minimize the effects of any damage.

Specially designed to measure the effects of meteoroid hits, NASA's

Long Duration Exposure Facility (LDEF) satellite orbited the Earth from April 1994 to January 1990. In its 69-month operational lifetime, LDEF suffered more than 15,000 small impacts on its hull – a small yet disconcerting proportion of these were caused by man-made space junk.

The thousands of satellites orbiting the Earth occupy a range of inclinations from polar to near-equatorial orbits, but they all cross the equatorial plane twice during each orbit – once heading north and again heading south. Consequently, the greatest chance of a mutual satellite collision occurs during equatorial plane passage. The effect of such a collision depends on the speed of the satellites relative to each other, and this in turn depends on the angle at which their orbits intersect at the equatorial plane. Satellites hitting each other head-on in a low Earth orbit (most likely if both satellites were polar orbiters, one heading north and the other heading south) would have a closing speed as high as 55,000km per hour. The effects of such a high-speed collision would be explosive and easily visible from the Earth's surface, even in daylight. However, satellites moving in close orbits in the same west–east direction and inclined only slightly to each other might just nudge against each other as one overtook the other.

In 1982, Czech astronomers L. Schnal and L. Popisilova calculated that a mutual spacecraft collision is likely once every 60 years, given a satellite population that remained at 1982 levels (some 4,500 objects). With space becoming ever more crowded, those odds are now reducing considerably. The planet-wide radar sensors of NORAD are constantly monitoring near-Earth space and can inform NASA on the threats posed to Shuttle missions by the 7,000 objects it has on its books, enabling the Shuttle to take evasive action if necessary. But there are estimated to be at least ten times the number of unseen small pieces of artificial debris capable of doing considerable damage to manned spacecraft. As an illustration of the dangers of debris around the Earth, in July 1983 one of the windows of the ill-fated Space Shuttle Challenger was pitted by a piece of space debris thought to have been nothing more vicious than a high-speed fragment of paint that had flaked off an old rocket booster.

In June 1997, the Russian space station Mir ('Peace') was involved in a potentially disastrous collision with an unmanned craft as it was about to dock. As it approached Mir, the small Progress M-34 resupply craft, containing fresh supplies for the 11-year-old space station, collided with the solar panels on the Spektr science module, causing extensive damage and decompression. There was a drop in power supply and the hull of the Spektr module was also damaged. Luckily, the situation was not serious enough to warrant an immediate order to 'abandon ship', although the cosmonauts were prepared to leave immediately if the situation worsened. This had been just a low-speed collision – it does not pay to imagine the awful consequences of the impact on Mir of a sizeable meteoroid travelling at high velocity.

METEOROID HAZARDS

Early in the history of space exploration, before the space-faring nations began to pollute space with their junk, meteoroids were regarded as the greatest hazard to manned and automatic spacecraft after the harsh vacuum of space itself. Scientists went to appreciable lengths to determine the numbers of meteoroids in near-Earth space by launching sounding rockets into the upper atmosphere and minutely examining the recovered experiments. The early results looked promising and suggested that the meteoroid risk to spaceprobes – at least, in terms of sheer number – was not as great as had been feared. In 1954, captured V-2 rockets were flown to heights of 150km over New Mexico, and meteoroid collisions were observed to take place at a height of around 40km. The impacts were detected by sensitive microphones which recorded the minute vibrations they produced, and detailed microscopic studies of recovered pieces of the rockets' fuselage showed the presence of tiny meteoroid impact pits.

In October 1957, the US Air Force conducted a successful experiment in which clouds of artificial meteoroidal material – actually, thousands of small ball bearings – were placed high above the Earth and their atmospheric re-entry and burn-up was observed. The ball-bearings were lifted skywards in the nose of an Aerobee rocket, which climbed to an altitude of 53km above the New Mexico desert. The nose cone then separated and coasted to a height of 86km (around the same altitude at which most meteors are observed to burn up) and an explosive charge was detonated to distribute the ball bearings high in the atmosphere. The explosion itself reached magnitude -10 (about the same brightness as a half Moon). The ball bearing meteoroids later re-entered and burned up in the atmosphere, producing meteors of about magnitude -4 (as bright as the planet Venus at its best), giving observers on the ground a thrilling fireworks display.

Ground-based experiments into the effects of meteoroid strikes included explosively accelerating particles of dust to high speeds (in a vacuum chamber to simulate the space environment) and firing the particles at suitable targets. At speeds of 30km per second – the fastest speed to which a particle could be accelerated explosively in 1960 – tiny particles were found to do as much damage on impact as the explosion of many times their own mass in TNT. This is about average velocity for Earth-bound meteoroids, which approach the Earth's atmosphere with speeds ranging from 12.8km to 72km per second, depending on whether the object is approaching Earth from 'behind' and overtaking our planet, or speeding towards the Earth in a head-on collision. Space vehicles in Earth orbit are moving targets and therefore may encounter particles travelling at a range of velocities extending slightly outside these parameters.

The chances of a meteoroid hit on an orbiting spacecraft were studied in 1948 by US scientist George Grimminger, who based his research around a hypothetical spacecraft with a surface area of 93sq m in a circular orbit at a height of 480km. In space, in the vicinity of the Earth, the average distance between sizeable meteoroids (those large enough to survive the Earth's atmosphere and be found as meteorites) is several hundred kilometres, and the risk of a spacecraft being damaged by one of these was considered extremely remote.

But for exposed vessels in space, even the smallest grains are capable of wreaking considerable damage, and it is these meteoroids that cause most concern. Grimminger calculated that every day the Earth is bombarded by 450,000 meteorites larger than 5mm in diameter. Smaller meteoroids are more numerous, with the daily arrival of some 45 million 1mm-diameter meteoroids in our planet's atmosphere. No fewer than 4.5 billion 200μ-diameter ($^1/_5$ mm size) micrometeoroids were thought to be swept up by the Earth's gravity day after day.

Grimminger's imaginary bombarded spacecraft was calculated to be hit by a 5mm meteoroid once every 38,800 years – about a two-million-to-one chance of such a hit in a single year. A 1mm-diameter meteoroid grain would impact on the vessel's hull every 388 years, and at least two 200-μ particles would hit every century. It is difficult to imagine even the most nervous astronaut being deterred by these figures.

Shooting at a Spacecraft

When a meteoroid hits the metal hull of a spacecraft, it will damage the surface at the point of impact. A small, slow-moving particle may simply dent the metal and bounce off into space. This happened during the Perseid meteor shower of August 1993, when cosmonauts aboard the Russian space station Mir, in orbit around the Earth at a height of about 360km, heard dozens of small meteoritic impacts on the hull. Later inspection during a space-walk brought to light several large holes in the fragile solar panel array.

If the meteoroid particle is going fast enough, and the hull is weak enough and fairly thin, the particle may penetrate completely and exit on the other side, leaving a hole barely larger than the particle that caused it, surrounded by a typical region of radial fractures. Such mechanical impacts by small meteoroids on spacecraft have been observed and recorded for decades. Usually, the spacecraft have survived such interplanetary sandblasting and the lives of astronauts have not been placed at significant risk. A small puncture hole in the pressurized living quarters of a manned spacecraft is unlikely to be followed by a major explosion, and the effects of a minor unobserved puncture might not even be noticed, since most spacecraft are not hermetically sealed and they leak into space anyway to some extent.

A larger particle will do both mechanical and thermal damage to a spacecraft at the point of impact. The impact of a 5mm meteoroid travelling at 76km per second will heat the impact site to a temperature exceeding 4,000,000°C; this will have the effect of a small explosion as the superheated metals vaporize and rapidly expand. If the impact happened against a pressurized liquid hydrogen or oxygen gas container or a fuel tank, the consequences would probably be disastrous. In 1975, the inflated balloon satellite called Pageos suddenly deflated when its pressurized hull was penetrated by such a high-speed object, possibly a meteoroid or a piece of anonymous space junk.

The American comet expert Fred Whipple was the first to suggest that the best way of protecting a spacecraft from high-energy meteoroidal impact is to install a 'bumper shield' around the most vulnerable parts of the hull. The shield, a thin metal plate composed of duralumin or stainless steel, would be positioned about 3cm from the hull. Any small impactor would hit the shield before hitting the spacecraft and lose most of its energy in the process. Should the impactor survive the shield and hit the spacecraft below, it might only dent the hull or produce a mechanical penetration hole which could be quickly sealed and later repaired. It was calculated that chrome steel would afford the best meteoroid protection – such a shield, 1mm thick with a surface area of 1,000sq m, would weigh about 800kg, and it would protect the spacecraft adequately from those impacts expected to occur during its operational lifetime in Earth orbit.

Probing the Meteoroids

In September 1957, the Russian space scientist Professor A. A. Blagonravov told a meeting of rocket and satellite engineers that future orbiting Soviet satellites would be used to gather important data about meteoroids in near-Earth space. He claimed that the probes would collect samples of micrometeoroids as they circled the Earth because this knowledge was vital for the purposes of interplanetary space travel. In November 1957, the Vice-President of the Soviet Academy of Sciences, I. P. Bardin, announced that the first orbiting satellite, Sputnik 1, had passed through several 'meteor showers' without any apparent damage.

Explorer 1, the US' first satellite, was launched atop a Jupiter C rocket at the end of January 1958, some four months after the Russian Sputnik 1 began to circle the Earth. The small satellite, a cylinder 2m long and 15cm in diameter, carried out four main experiments which measured the probe's external skin temperature, its internal temperature, cosmic ray data and meteoroid erosion; the latter experiment was conducted by the US Air Force Research Center. With its sensitive microphone, Explorer measured the rates and density of meteoroids impacting on the probe's outer shell. Installed at the probe's aft end, a parallel set of 12 narrow wires were meant to detect the arrival of any dense shoals of micrometeoroids – if a wire happened to be severed by an impact, the change in current would be detected. In the end, Explorer did not encounter large clouds of sandblasting meteoroids, but a number of individual meteoroid hits were recorded.

In October 1958, the US Moon probe Pioneer 1B was launched. Ultimately, Pioneer's Moon mission proved to be a failure because the probe did not reach an adequate velocity to escape from the Earth; instead, the probe climbed to an altitude of 114,000km and fell back to Earth. However, during the first nine hours of the probe's ascent, a sensitive meteoroid microphone experiment detected the impact of 11 small, slow-moving meteoroids. A month later, a second Moon probe called Pioneer 2 detected a veritable mini-shower of meteoroids with its microphone experiment. The first meteoroid impact was felt at a height of over 1,100km, followed by rates of about 16 per minute at 1,400km. Like its predecessor, this probe's Moon mission was never realized, but at least valuable data on the space environment was secured.

In 1963, the US' Explorer XVI, with its meteoroid-sensitive surface area of 2.3sq m, obtained information on the size and velocity of small meteoroids. Two years later, in February, May and July 1965, three huge meteoroid detection satellites called Pegasus were launched atop three Saturn I rockets. The meteoroid detectors were attached to the upper stages of the Saturn rockets and deployed in Earth orbit. Each Pegasus satellite weighed 2 tonnes, and consisted of a 3m-wide concertina panel which unfolded to a span of 29m once in orbit.

From orbital heights ranging from 500km to 1,300km, the Pegasus probes made valuable observations of the population of meteoroids in near-Earth space. Pegasus' giant 'wings' were a sandwiched construction of aluminium, Mylar and copper through which a small electric current flowed. The impact of a small meteoroid upon the panel gave rise to a small conductive gas bubble which temporarily short-circuited the equipment, discharging the capacitor. The resulting electrical pulse was recorded as a strike and the information was telemetered back to ground stations on Earth. By the summer of 1966, the Pegasus probes, with their combined panel area of more than 500sq m, had detected more than 1,100 small meteoroid impacts, amounting to two strikes per square metre. None of these impact events, however, was deemed to have been dangerous enough to have destroyed any of the manned Apollo missions to the Moon.

Perseid Perils

Because of the possible risks associated with meteoritic impact, NASA postponed the launch of the space shuttle Discovery until after the predicted high maximum of the Perseid meteor shower in August 1993. Fearing the effects of a prolonged meteoritic strafing, NASA also wisely slewed their precious space observatories in Earth orbit – the Hubble Space Telescope, the Compton Gamma Ray Observatory and the

Extreme Ultraviolet Observer – to face a direction opposite the Perseid radiant, in order to minimize damage. Prior to the event, it had been calculated that an object the size of the Hubble Space Telescope has one chance in 1,000 of being hit by a sizeable Perseid meteoroid during the space of a 15-minute meteoroid storm – the same chance as of it being blasted to pieces by a 1m-diameter meteoroid over a period of 17 years.

When 1993's Perseids did arrive, cosmonauts aboard the Russian space station Mir reported hearing the unnerving pitter-patter of meteoritic impacts on the hull of their spacecraft. Later inspection during three space-walks revealed that sections of Mir's fragile solar panels had large holes in them, one the size of a football! A Perseid meteoroid impact with the ESA's direct-broadcast satellite Olympus (cost: $850 million) during the shower maximum appears to have knocked it out of control, and the subsequent battle to stabilize the craft depleted most of its altitude control propellants.

Perils Further Afield

Early space pioneers saw meteoroids as a potential hazard to manned lunar exploration. According to one 1960s' view (quoted by Jack Coggins and Fletcher Pratt – see Bibliography), 'the suit for work on the Moon will not be a suit at all, but something like a small army tank!... the original rocket, of course, will have sides that are stout and thick enough to resist any meteorite it is likely to meet.'

Micrometeoroids were later downgraded in the risk-to-safety stakes. The Apollo spacesuit, an 'integrated thermal micrometeoroid garment' with layers of strong sandwiched synthetic material, was fairly comfortable and flexible, giving a degree of protection if the astronaut happened to slip on to a jagged lunar rock or was hit by a small meteoroid. Apollo's lunar lander was a very flimsy affair, however, and despite its apparent solidity it had paper-thin metal walls which, during their exposure to the vacuum of space, were pockmarked by a multitude of tiny meteoroid impacts. Analysis of the external window surfaces of the Apollo command modules (which were the only large pieces of Apollo hardware to return to earth) has revealed the presence of tiny micro-craters melted into the glass, each with its own little collar of melt material.

Meteoroidal Martian Mishaps

The US' Mars probes, Mariner 6 and 7, were launched in early 1969. In July, as both craft neared the red planet, the deep-space tracking station at Johannesburg in South Africa was disconcerted to find that it had lost Mariner 7's radio signal. Some hours later, when the probe had 'risen' in the skies over the Pioneer station at Goldstone, a very weak signal was found. Fearing that half of the $148 million Mariner mission might be at risk, commands were quickly sent for Mariner 7 to use an alternative antenna. Eleven minutes later – the light-speed time taken for the command to reach the probe and for the new signals to be relayed back to Earth – relieved engineers at the Goldstone and Australian stations reported a good strong signal. It was thought that Mariner 7 had been bumped by a chance meteoroid hit and that as a result the probe had temporarily lost its optical lock on the star Canopus – a bearing the probe's computer needed in order to orientate itself properly in space.

The asteroid belt lies beyond the orbit of Mars, between 1.7AU and 4AU from the Sun. The zone of minor planets probably contains more than a billion asteroids larger than 1km in diameter, although only 5,000 of the larger asteroids have so far been identified and their orbits calculated. The asteroid Ceres, a little larger than 900km across, is the largest member of the family of minor planets. The total mass of the asteroids probably amounts to some 60 million million tonnes, less than one-thousandth of the mass of the Earth. Popular science fiction would have us believe that asteroid belts are pretty hazardous places, with barely room to fit a spacecraft between the individual members. In fact, the average distance between asteroids larger than 1km in diameter is so great that your nearest asteroid neighbour would appear only as a feeble point of light. The chances of a spaceprobe colliding with an asteroid as it heads for the outer solar system are so remote as to be negligible.

Now that it seems as though we humans are actually going to make it into the new millennium without having annihilated ourselves in some nuclear-heated squabble over political or religious differences, the future looks promising. Near-Earth space will continue to be exploited by satellites, manned and unmanned, and it will not be long before permanently occupied space stations are established in Earth orbit, on the Moon and on Mars. Over the next couple of decades, the danger of asteroidal or cometary collisions with our planet must be seriously addressed. Ultimately, there needs to be a reliable system of long-range detection for the larger objects, and the capability of destroying or diverting them before they get anywhere near the Earth. As long as the Earth has an atmosphere, the smallest bits of cosmic debris will always evade detection, until they make their presence known by burning up and shining in the sky as beautiful shooting stars.

USEFUL ADDRESSESS

OBSERVING METEORS AND REPORTING FIREBALLS

Society for Popular Astronomy (SPA)
Secretary: Guy Fennimore
36 Fairway
Keyworth
Nottingham NG12 5DU
England

SPA Meteor Section Director Alastair McBeath informs members on how to observe meteors and contribute to the SPA observing programme. The SPA Meteor Section has close ties with the International Meteor Organization (IMO).

International Leonid Watch
Contact: Paul Roggemans
Pijnboomstraat 25
B-2800 Mechelen
Belgium

Run under the auspices of the IMO.

IMO Fireball Data Centre (FIDAC)
c/o Andre Knoefel
SAARBRÜCKERSTR 8
D-W-4000
Düsseldorf
Germany

British Astronomical Association
Co-ordinator: Howard Miles
Lane Park
Pityme
St Minver
Wadebridge
Cornwall PL27 6PN
England

Satellite observation; also reports of fireballs and other unusual aerial phenomena.

Spaceguard UK
35 Pownall Road
Larkhill
Salisbury
Wiltshire SP4 8LX
England

Affiliated to the international Spaceguard Foundation, Spaceguard UK was established to pursue the following aims:

• To promote and encourage British activities involving the discovery and follow-up observations of Near-Earth Objects.
• To promote the study of the physical and dynamic properties of

asteroids and comets, with particular emphasis on Near-Earth Objects.

• To promote the establishment of an international, ground-based surveillance network (the Spaceguard Project) for the discovery, observation and follow-up study of Near-Earth Objects.

• To provide a national United Kingdom information service to raise public awareness of the Near-Earth Object threat, and technology available to predict and avoid dangerous impacts.

METEORITES

Good general museum displays containing all the main types of meteorite can be viewed at the venues listed below.

British Museum of Natural History
Cromwell Road
South Kensington
London SW7 5BD
England

American Museum of Natural History (incorporating the Hayden Planetarium)
Central Park West at 77th Street
New York
NY 10024
USA

On display is the famous Ahnighito iron meteorite (31 tonnes) which was found at Cape York in Greenland and shipped south by Robert Peary in 1897. Also on show is the 15 tonne Wilamette meteorite (which fell at Wilamette, Oregon).

Department of Meteorites NHB119
Smithsonian Institution
14th and Constitution Avenue NW
Washington DC 20560
USA

The Geological Survey of Canada
Meteorite Identification Department
601 Booth Street
Ottawa
Ontario
Canada K1A OE8

Offers a $500 (Canadian dollar) reward for every new Canadian meteorite find.

School of Mines
Mexico City

On display is the 20 tonne Bacbirito meteorite which was found at Sinaloa, and the 21 tonne Chupaderos meteorite (two fragments) found at Chihuahua, Mexico.

The largest single meteorite ever to be recovered, the 60 tonne Hoba West specimen, is on display where it fell in prehistoric times at Grootfontein in Namibia.

BIBLIOGRAPHY

Abbreviations

AN – *Astronomy Now* magazine
BAA – British Astronomical Association
BAS – Birmingham Astronomical Society
BIS – British Interplanetary Society
JBAA – Journal of the BAA
JBIS – Journal of the BIS
JASNC – JAS News Circular
PA – Popular Astronomy, quarterly magazine
 of Britain's Society for Popular Astronomy
S&T – Sky & Telescope magazine
SPA – Society for Popular Astronomy
 (formerly JAS – Junior Astronomical Society)
Spaceflight – magazine of the BIS
SPANC – SPA News Circular

Introduction

Berry, Richard, and Burnham, Robert, 'Voyager 2 at Saturn', *Astronomy*, November 1981

1 Comet Tales

Berry, Richard, 'Giotto Encounters Comet Halley', *Astronomy*, June 1986
Berry, Richard, and Talcott, Richard, 'What Have We Learned From Comet Halley?', *Astronomy*, September 1986
Grego, Peter, 'Four Comets of the 1980s', BAS Newsletter, March 1990
Larousse Encyclopedia of Astronomy (1959)
Marsden, Brian, 'The Kohoutek Controversy', *Hermes*, October 1974
O'Meara, Stephen James, 'Probing Comets with Meteor Streams', *S&T*, June 1995 (based on an article in *Icarus*, June 1994)
Ridpath, Ian, 'Dusty Trails Make Meteor Showers', *PA*, October 1986
Sutherland, Paul, 'Levy Puts on a Good Show', JASNC 159, September 1990

— 'Halley's Comet in Mystery Flare-up', JASNC 162, March 1991
— 'Shoemaker-Levy – A Binocular Comet', JASNC 169, May 1992
— 'Gotcha! Swift-Tuttle is Tracked Down at Last', JASNC 172, November 1992
— 'New Monster Comet Riddle', SPANC 187, August 1995
Yeomans, Donald, *Comets* (Wiley, 1991)

2 Meteors: Lights in the Sky

Edberg, Stephen, and Levy, David, *Observing Comets, Asteroids, Meteors and the Zodiacal Light* (Cambridge University Press, 1994)
McBeath, Alastair, 'Observing Shower Meteors', *PA*, July 1988, January 1989 and October 1989
— 'Sporadic Meteor Colors' and 'Shower Meteor Colors', *WGN*, Journal of the IMO, 18,4 (1990) and 19,5 (1991)
Monthly Notices of the Royal Astronomical Society, June 1993 (University of Western Ontario astronomer Martin Beech is quoted in SPANC 175)

Ridpath, Ian, 'Parent of the Quadrantid Meteors', *AN*, December 1993

Stephens, Paul, 'The Aurora of 1991 Nov 8/9', BAS Newsletter, October 1991

Stoy, R. H., *Everyman's Astronomy – The Debris of the Solar System* (J. M. Dent & Sons Ltd, 1974)

Union Observatory Circular, Nos 1 to 44 , 1 October 1912 to 17 January 1919 (Pretoria, SA Govt Printing & Stationery Office, 1919) (No 1 reports a remarkable daylight meteor, and contains a photograph and diagram of the trail it left in the sky)

3 The Lion Roars Tonight

Burke, John G., *Cosmic Debris (Meteorites in History)* (University of California Press, 1986)

Chambers, G. F., *A Handbook of Descriptive and Practical Astronomy* (Clarendon Press, 1889)

Ellicott, Andrew, 'Account of an Extraordinary Flight of Meteors (Commonly Called Shooting Stars)', *Transactions of the American Pholosophical Society*, No 6, 1804

Hawks, Ellison, *Stars* (T. C. & E. C. Jack, 1910)

Mason, John, 'The Leonid Meteors and Comet 55P/Tempel-Tuttle', JBAA, Vol 105, No 5, October 1995

Newton, Henry, 'The Original Accounts of the Displays in Former Times of the November Star Shower', *The American Journal of the Sciences and Arts*, No 37, 1864

Olivier, Charles, *Meteors* (Williams and Wilkes, 1925)

Olmsted, Prof Denison, 'Observations of the Meteors of 13 November 1833', *American Journal of the Sciences and Arts*, Nos 25 and 26, 1834

Rao, Joe, 'The Leonids: King of the Meteor Showers', *S&T*, November 1995

Von Humboldt, Alexander, *A Personal Narrative of Travels to the Equinoctial Regions of the New Continent* Vol 3 (Longman, 1818)

4 Meteorites: Souvenirs from Space

Davidson, Martin, *Astronomy for Everyman* (J. M. Dent & Sons Ltd, 1954)

Graham, Andrew, et al., *Catalogue of Meteorites* (University of Arizona Press, 1985)

'A New Piece of the Moon', *S&T*, October 1994

Proctor, Richard, 'The Gemsbroke Incident', *Knowledge*, 2 February 1883

S&T, March 1995 (p. 16: meteor fatality recorded in ancient Chinese accounts; also sixteenth-century woodcut of meteorite fall)

5 Cosmic Collisions

Anderton, Colin, 'One Giant Leap for Brian', BAS Newsletter, Vol 14, No 3, December 1989

— 'Minor Planet 4751 Alicemanning', BAS Newsletter, Vol 15, No 6, June 1991

'Another Car Conker', *S&T*, September 1995

Ashbrook, Joseph, 'Lunar Meteoroid Showers', *S&T*, August 1976

'Astroblemes,' *Scientific American*, August 1961

Barnes-Svarney, Patricia, 'The Near-Earth Asteroids', *PA*, July 1989

Beatty, J. Kelly, and Goldman, Stuart, 'The Great Crash of 1994: A First Report', *S&T*, October 1994

Berry, Richard, 'Earth Gains a Neighbour', *Astronomy*, August 1989

Bevan, A. W. R., and Hutchinson, R., *Catalogue of Meteorites* (Natural History Museum), 1985)

Bibliography of Terrestrial Impact Structures (NASA, 1985)

Blaylock, Kevin, and McBeath, Alastair, 'Meteor Marathon', *PA*, April 1985

Bone, Neil, 'Catch a Falling Star – on Film!', *PA*, October 1985

Burke, John G., *Cosmic Debris (Meteorites in History)* (University of California Press, 1986)

Burnham, Robert (Ed), 'Did an Asteroid Create the Everglades?', *Astronomy*, May 1986

Capen, Charles, 'Hunting Martian Astroblemes', *Astronomy*, September 1986

Coggins, Jack, and Pratt, Fletcher, *By Spaceship to the Moon* (Publicity Products Ltd, *c*.1960)

Dickey, Beth, 'NASA Launches Meteor Watch', *AN*, December 1993

Dietz, Robert S., entry for Meteorite Craters in *Grolier Multimedia Encyclopedia 1996*

Furneaux, Rupert, *The Tungus Event* (Granada, 1977)

Gallant, René, *Bombarded Earth* (John Baker, 1964)

Gallant, Roy, 'Journey to Tunguska', *S&T*, June 1994

Gordon, Bonnie, 'Jupiter Took its Lumps in 1690', *Astronomy*, May 1997

Hosking, Tom, 'Identifying Threats to Earth', *PA*, October 1994

Miles, Howard, 'The Barwell Meteorite and Associated Fireballs', *JBAA*, Vol 77, No 3, April 1967

Lewis, Roy, et al., 'Diamonds in the Sky', *PA*, July 1987 (from *Nature*, Vol 326, March–April 1987)

McBeath, Alastair, 'Sporadic Meteors', *PA*, July 1985

— 'How to Observe Fireballs', *PA*, January 1992

— 'The Tears of St Lawrence', *PA*, July 1992

Meredith, Cliff, 'The Jupiter Event in Context', *PA*, July 1994

'New York's Cosmic Car Conker', *S&T*, February 1993

O'Meara, Stephen James, 'The Great Dark Spots', *S&T*, November 1994

Ridpath, Ian, 'Comet Encounters Have Changed the World', *PA*, October 1982

— 'Earth's Orbiting Scrapyard', *PA*, April 1986

— 'Mind Your Head!', *PA*, April 1986 (from *Nature*, Vol 318, November 1985)

Rogers, John, 'The Comet Collision with Jupiter: I. What Happened in the Impacts', *JBAA*, Vol 106, No 2, 1996

Spalding, George, 'The Tears of St Lawrence', *Hermes*, July 1980

Stephens, Paul, 'The Christmas Meteorite', BAS Journal, February 1996

Sutherland, Paul, 'Minor Planet Man Brian's Number Is Up', *JASNC* 158, July 1990

Watson, Roy, 'To Catch a Falling Star', *PA*, April 1995

Yeomans, Donald, *Comets* (Wiley, 1991)

6 Watching the Skies

Bone, Neil, 'Catch a Falling Star – on Film!', *PA*, October 1985

Eberst, Russell, 'You Can Spot a Satellite', *PA*, January 1987

Edberg, Stephen, and Levy, David, *Observing Comets, Asteroids, Meteors and the Zodiacal Light* (Cambridge University Press, 1994)

Flammarion, Camille, *Astronomy for Amateurs* (Nelson, 1903)

7 Famous Fireballs, UFOs and Meteor Miscellany

'Asteroids Ahoy!', *Fortean Times* 97, April 1997 (reported from the *Daily Telegraph*, 16 October 1996)

Baker, David, *The History of Manned Spaceflight*

Bond, Peter, 'Snowballs Strike the Earth', *AN*, July 1997

Canby, Courtlandt, *A History of Rockets and Space* (Leisure Arts Ltd, 1962)

Cleator, P. E., *Into Space* (Allen & Unwin Ltd, 1953)

Coggins, Jack, and Pratt, Fletcher, *By Spaceship to the Moon* (Publicity Products Ltd, c.1960)

Collins, Michael, *Carrying the Fire* (W. H. Allen, 1974)

Edberg, Stephen, and Levy, David, *Observing Comets, Asteroids, Meteors and the Zodiacal Light* (Cambridge University Press, 1994)

Fairley, Peter, *Man on the Moon* (Arthur Barker Ltd, 1969)

Gatland, Kenneth, 'Large Meteoroid Satellite', *Spaceflight*, Vol 5, No 5, September 1963

—'Cosmic Debris in Lunar Orbit', *Spaceflight*, Vol 8, No 8, August 1966

—'Punctured Pegasus', *Spaceflight*, Vol 8, No 10, October 1966

— *Manned Spacecraft* (Blandford Press, 1967)

— *Robot Explorers* (Blandford Press, 1972)

Good, Timothy, *Above Top Secret* (Grafton Books, 1989)

Grego, Peter, 'The UFO Files of W. E. Marsh', BAS Newsletter, April 1993

Grimminger, George, 'Probability that a Meteorite will Hit or Penetrate a Body Situated in the Vicinity of the Earth', *Journal of Applied Physics*, Vol 19, 1948

Hough, Peter, and Randles, Jenny, *The Complete Book of UFOs* (Piatkus, 1994)

'Icebergs from Space', *Fortean Times* 102, September 1997

Kostko, Oleg, 'Looking for Cosmic Microbes', *Spaceflight*, Vol 6, No 5, September 1964

Langton, N. H., 'Meteors and Space-flight', *Spaceflight*, Vol 1, No 3, April 1957

Mallan, Lloyd, *Space Satellites* (Fawcett, 1958)

'A Meteorite That Missed Earth', *S&T*, July 1974

Monthly Notices of the Royal Astronomical Society, June 1993 (University of Western Ontario

astronomer Martin Beech is quoted in SPANC 175)

Moorcroft, Norman, 'Space Exploration', *Hermes*, April 1978

Murdin, Paul, and Allen, David, *Catalogue of the Universe* (Cambridge University Press, 1979)

Pollard, Frank, 'US Satellites (Failure and Success), Artificial Meteors and Project Farside', *Spaceflight*, Vol 1, No 7, April 1958

Ridpath, Ian, 'It's Hissing Syd(ney)!', *PA*, January 1981

— 'A Crash Course on Space Traffic', *PA*, July 1982

— 'Earth's Orbiting Scrapyard', *PA*, April 1986

—'Parent of the Quadrantid Meteors', *AN*, December 1993

Robey, Donald H., 'Meteoritic Dust and Ground Simulation of Impact on Space Vehicles', JBIS, Vol 17, No 1, January 1959

— 'An Hypothesis on the Slow-Moving Green Fireballs', JBIS No 17, 11 September 1960

Roberts, J. G., 'The Artificial Comet', *Spaceflight*, Vol 5, No 6, November 1963

Sachs, Margaret, *The UFO Encyclopedia* (Corgi, 1980)

Salmon, Andy, 'Human Meteors', *PA*, July 1997

Sutherland, Paul, 'Mystery of the Nineteenth Hole', *Hermes*, April 1979

Van Flandern, Tom, 'The Cyrillid Meteors', Meta Research Bulletin 1, pp 32–3, 1992

White, C. J., and Blackburn, P. B., *The Elements of Theoretical and Descriptive Astronomy* (John Wiley & Sons, 1920)

White, Dale, *Is Something Up There?* (Doubleday, 1968)

INDEX

Page numbers in *italics* refer to figures.